Gallus Thomann

Real and Imaginary Effects of Intemperance

Gallus Thomann

Real and Imaginary Effects of Intemperance

ISBN/EAN: 9783337371432

Printed in Europe, USA, Canada, Australia, Japan

Cover: Foto ©berggeist007 / pixelio.de

More available books at **www.hansebooks.com**

REAL AND IMAGINARY

EFFECTS OF INTEMPERANCE.

A STATISTICAL SKETCH,

CONTAINING LETTERS AND STATEMENTS FROM THE SUPERINTENDENTS OF EIGHTY AMERICAN INSANE ASYLUMS, THE HISTORY OF FIVE HUNDRED INEBRIATES, THE HISTORY OF SIX HUNDRED AND SEVENTY-ONE PAUPERS, AND STATISTICS OF DRUNKENNESS; TOGETHER WITH A REVIEW OF THE OPERATIONS OF PROHIBITORY AND RESTRICTIVE LAWS, AND THE GOTHENBURG SYSTEM.

BY

G. THOMANN.

NEW YORK:
THE UNITED STATES BREWERS' ASSOCIATION.

1884.

TO THE PUBLIC:

We submit these facts, figures and arguments on real and imaginary effects of intemperance with a just estimate of the prejudice they are sure to encounter from the foregone conclusion that they are published, not for the advancement of truth, but in the interest of the brewing trade. It would be worse than useless to claim that our motives are free from self-interest; we ask credit, however, for an honest endeavor to offer in this treatise authentic information on a subject, concerning which much has been published which rests on conjecture merely. Our plain statements may offend many whose good-will we value in our business relations, and this one fact will, perhaps, serve better to gauge our sincerity, than any protestations we could make.

THE UNITED STATES BREWERS' ASSOCIATION.

NEW YORK, May, 1884.

ERRATA.

Page 5, line before last of foot-note, read *three* instead of *four*.

Page 26, No. 52 of table, 3rd, 4th and 5th columns, read : 1—0—1. instead of 15—8—23.

CONTENTS.

CONTENTS—*Continued.*

REAL AND IMAGINARY

EFFECTS OF INTEMPERANCE.

INSANITY.

The advocates of prohibition have always manifested a strong aversion to statistics; not only have they neglected to exercise their talents in the study of this important science, but they have treated even such incontestable facts as appear from the reports of the Internal Revenue Office with uncommon fastidiousness and incredulity. As a rule, they have given preference to sentimental arguments which, while they admit of a very high order of rhetoric and an almost limitless scope for imaginative embellishments, appeal to the heart rather than to the reason. The very nature of their position has necessitated such a course. Honest facts are often most concisely expressed by figures, but sentiments of a certain character require the gaudy garb of striking phrases; and such phrases, whether they be pregnant with wisdom or utterly devoid of it, exercise great influence upon people who are either predisposed to uphold the sentiments expressed, or indifferent to their truthfulness.

A French writer recently asserted, and proved, to his satisfaction at least, that the French nation was governed by witticism. To any one acquainted with French history, the grounds upon which this somewhat hyperbolic conclusion is based, must be obvious enough. By analagous reasoning it might, in the light of our political history, be made to appear that, when no vital questions are at stake, American voters are easily led into most singular actions by phrases pure and simple, provided that these convey a pleasant and "catching" sentiment.* Prohibitionists never had facts on their side, but they have had, and have, an abundance of theories and a world of sentiment at their command. They have conceived the grand idea that an ideal social state is not an impossibility in this world of imperfections—and they have found a host of believers.

Following such guidance, the victims of their eloquence labored under constant illusions and delusions, eagerly pursuing the improbable, and leaving truth far behind them; so that they might not inappropriately be likened to the visionaries and theorists, whom Goethe has satirically characterized:

> "I say to thee, a speculative wight
> Is like a beast on moorlands lean,
> That round and round some fiend misleads to evil plight,
> While all around be pastures fresh and green."

In their arguments the leaders of this agitation have cared neither for primary causes nor general results, but have looked only to effects in individual cases, building upon such isolated experiences curative theories for all the evils of the universe.

Of late, however, these deserving persons who, solely guided by motives of philanthropy, persist in devoting much of their time and energy to the thankless task of ameliorating the condition of their fellow-creatures, are beginning to pay considerable attention to figures. They have, to some degree, deserted the "moorlands lean" and betaken themselves to the "pastures fresh and green" of statistics. A number of amusing attempts at statistical argumentation would seem to indicate that they regard the census of 1880 as a veritable magazine of formidable weapons, with which to annihilate the arguments of their adversaries of bibulous propensity.

Indeed this census, perfect as it is in most respects, imperfect as, in the very nature of things, it cannot but be in others, bids fair to become the future basis of prohibitory arguments—not so much on account of what it actually presents, as because of the inferences and deductions to which, under certain aspects, it may be made to lend a coloring of plausibility.

* "Betty and the Baby" is one of the more recent epigrammatic productions of this sort. Dr. Von Holst, in his Constitutional History of the United States, tersely characterizes some of the senseless battle cries of political campaigns, by which, without any reasonable grounds either for enthusiasm or indignation, popular sentiment is often stimulated to an exalted pitch in either direction. The same author quotes from a letter of A. Hamilton to J. A. Bayard: "Nothing is more fallacious than to expect to produce any valuable or permanent results in political projects by merely relying on the reason of men. Men are rather reasonable than reasoning animals, for the most part governed by passions."

It is well known that, viewed from a prohibitory standpoint, insanity, pauperism and crime are *principally* attributable to inebriety, either directly or indirectly. According to some writers, the proportion of insanity caused by intemperance directly is from fifty to seventy per cent., while according to others, it is only from thirty-five to fifty. Insanity transmitted from drunken parents to their offsprings is said to vary from fifty to seventy-five per cent.,* and this ratio is alleged to be incessantly increasing. While dwelling with epic circumstantiality upon the horrors of inebriety, temperance advocates and prohibitionists formerly made no attempt to prove such assertions, which, in truth, they only used to point the moral of their tearful tales, relying for the rest upon the credulity of the timorous, and snapping their fingers at statistical criticism.

Now, however, they pretend to be able to adduce, if not conclusive proof of, at least corroborative data for, the correctness of their averments; and it is for this purpose that the census is pressed into their service.

INCREASE OF INSANITY.

The mode of reasoning by which temperance advocates arrive at their gloomy conclusions, is singularly arbitrary.

From the census of 1880 it appears that there were in the United States at that time 91,997 insane persons, against 37,432 in 1870. The increase of insane within the decade amounts to one hundred and forty-five per cent.; while that of the population only amounts to about thirty-three per cent. This glaring disproportion must in some manner be explicable; and what explanation could be less brain-taxing, more convenient and at the same time more suggestive in the eyes of prohibitionists, than that which charges the excess to increasing inebriety? None, of course; hence this explanation is adopted. But never was *a posteriori* reasoning more ludicrously illogical than in this case. It practically amounts to the following declaration: "Insanity, pauperism and crime develop in exact proportion as inebriety increases; hence whenever an augmentation of either of these three evils is manifest, there must also

* Dr. Bascom, President of Wisconsin State University, in his Philosophy of Prohibition says: "Of three hundred idiots in Massachusetts, Dr. Howe referred one hundred and forty-five to intemperance directly. (sic!) A like proportion of insanity finds a similar reference." What may be a perfectly correct statement of facts on the part of Dr. Howe is here tacitly made the premise of a very venturesome conclusion by Dr. Bascom, for in no part of his book is there an intimation that this proportion, founded upon an isolated inquiry, should not be taken as a general basis, and thus the reader is led to believe, that what is true of four hundred idiots in Massachusetts, applies also to the entire insane population of the United States.

be clearly discernible in corresponding ratio an increase of inebriety. Insanity has increased one hundred per cent. between 1870 and 1880, hence inebriety must have increased in like proportion."

An illustration may serve to bring the inconsequence between premise and conclusion into bolder relief. It is perfectly logical, for instance, to maintain that because the streets were sprinkled they must be wet; but what would be thought of a person who should contend that because the streets are wet, they must necessarily have been sprinkled?

With certain reservations as to the kind of drink used, the climate, the temperament and habits of the drinkers, as well as their social condition, it may readily be conceded that an increase of inebriety implies a proportionate increase of insanity; but it is not, therefore, by any means fair to infer an increase of inebriety from an alleged increase of insanity. The former proposition is almost axiomatic; the latter, putting it very mildly, is absurd, since its adoption as a true statement of fact would necessitate the exclusion of all the other multiform causes of insanity, as well as involve an abandonment of all the objections which have been and will again be urged by specialists, who take issue with the census officials in more than one respect.

Before approaching the subject of intemperance as a cause of insanity, let us consider whether the reported increase of lunacy is real or, to some extent at least, only apparent. If it be real, then, of course, it must be assumed that the census of 1870 contains a complete and wholly correct enumeration of the insane population in the country. But this assumption has been, and is still being, very vigorously assailed by eminent authorities. The fact that the Superintendent of the Census deemed it necessary, in 1870, to vindicate the correctness of his statistics of insanity must be taken as sufficient evidence of the reasonableness of doubt—a doubt, however, which does not reflect on the efficiency and conscientious diligence of the census officials.* The objections urged against accepting, as wholly reliable, the insanity statistics of 1870 are worthy of brief review.

Dr. E. T. Wilkins, in his report on "Insanity and Insane Asylums," a work frequently referred to in laudatory terms, and copiously quoted by medical writers, says: "It is exceedingly interesting to trace this apparent increase of insanity in various countries of the world, and easy to show how much more rapidly the increase has been brought to light in those countries where the

* Census of the United States, vol. II, p. 425.

most humane and liberal provisions have been made for their accommodation. We will show, however, that it is not confined to the present epoch, nor to any particular country; but that it has always and everywhere come forth from its recesses and hiding places whenever suitable hospitals for the reception of its victims were provided. Thus Bucknill and Tuke state that in the short space of nineteen years the estimated proportion of the insane in England rose from one in seven thousand two hundred to one in seven hundred and sixty-nine, while on the 1st of January, 1871, there was one to four hundred." *

An imposing array of data from other European countries as well as from the United States is then presented by Dr. Wilkins, all strongly corroborating the above assertion.

It need scarcely be affirmed that this increase appeared in Spain, a country noted for the sobriety of its people, quite as strikingly as in Denmark, where the people are addicted to the excessive use of ardent spirits. Dr. Jarvis, in a paper referred to in Dr. Wilkins' report, expresses his opinion in the following words:

"It will readily be supposed that the opening of new establishments for the cure and protection of lunatics, the spread of their reports, the extension of the knowledge of their character, power and usefulness by the means of the patients that they protect and cure, have created and coutinue to create more and more interest in the subject of insanity and more confidence in its curability. Consequently, more and more persons and families who kept their insane relations at home, now believe that they can be restored or improved, and therefore, send them to these asylums and thus swell the lists of their inmates."

As an argument in favor of the assumption that the increase of insanity, as shown by census reports, is to a great extent apparent, not real, the opinion of Dr. Jarvis seems well founded, even if more than due weight be given to the fact, that the enumeration, conducted by the Census Bureau, was not coufined to the insane in institutions, but comprised all demented persons *wherever found*. The difficulty lies in finding those insane who are treated at home, if treated at all. The want of adequate accommodation for the insane in any locality is in itself an evidence of a lack of interest in the matter on the part of the public, and it is reasonable to presume that in such localities all prejudices, not to say superstitions,

* Wilkins "Insanity, etc.," p. 57, (Sacramento, Cal., 1871.)

with which this terrible malady is sometimes regarded, prompting sensitive people to a concealment or denial of the presence of insanity in their family, still prevail to a greater or less extent.

The following table compiled from material contained in the census of 1880, shows a much greater proportion of insane in States where accommodations for the insane are ample:

STATE.	POPULATION	Numb'r of Insane Hospitals.	INSANE POPULATION.		
			Total Number of Insane.	Number of Insane in Hospitals.	Number of Insane at Home.
Alabama	1,262,505	1	1,521	373	1,050
Florida..............	269,493	1	253	76	175
Georgia...............	1,542,180	1	1,697	626	1,036
Kentucky...........	1,648,690	3	2,784	1,404	1,309
Louisiana	939,946	3	1,002	450	534
Mississippi..........	1,131,597	1	1,147	387	713
North Carolina......	1,399,750	1	2,028	269	1,518
South Carolina.	995,577	2	1,112	425	662
Tennessee	1,542,359	1	2,404	385	1,757
Texas.............. ..	1,591,749	1	1,564	350	1,119
Virginia............	1,512,565	4	2,411	1,098	1,123
	13,836,411	19	17,923	5,843	11,026
New York..........	5,082,871	27	14,111	8,079	4,421*

* The number of insane in jails, almshouses, etc., is not included in the above table.

From this table it will be seen that according to the census of 1880, eleven Southern States with a population of nearly fourteen millions, and nineteen insane hospitals, have an insane population of seventeen thousand nine hundred and twenty-three, while the State of New York, with only five millions of inhabitants and twenty-seven insane asylums, has over fourteen thousand persons thus afflicted. Of these, eight thousand and seventy-nine are in asylums and only four thousand four hundred and twenty-one at large, while in the eleven Southern States, the proportion is almost exactly reversed.

It is clear, of course, that innumerable circumstances, unknown in the Southern States, combine to augment the list of insane in New York. But making allowance for all these—for the large influx of foreigners; for the floating population which the State has to care for; for the immense proletariate of the metropolis; for the terrible effects of want and privation; the consequences of factory life; the extraordinary mental and bodily strain, to which all are subjected who take part in the feverish activity of the commercial and industrial centre of the land; for all the vices, whose rank growth is inseparable from the development of large cities—yet, there still remains a very large difference between the ratio of insanity in New York and in the eleven Southern States, which can only be accounted for in the manner already described. Upon comparing Massachusetts with Alabama, Kentucky and Tennessee, it will be found that the former State, with one million seven hundred and eighty-three thousand and eighty-five inhabitants and *fifteen* asylums, has five thousand one hundred and twenty-seven insane, of whom three thousand and eighty-five are in asylums, four hundred and eighty-two in almshouses and other institutions and fifteen hundred and sixty at home; while the three Southern States with four millions four hundred and fifty-three thousand seven hundred and ninety-two inhabitants, and five asylums, have in the aggregate six thousand seven hundred and nine insane persons, or fifteen hundred more than Massachusetts.

If it be kept in view that the movement in favor of asylum and prison reforms, inaugurated by a few humanitarians, and helped forward by eminent English, German and French writers, did not assume a very promising aspect, until about the fourth decade of this century, and that since then popular feeling, aided and sustained by scientific efforts and governmental measures, has brought about an entire revolution in this respect, it will not be wondered at that the censuses of the past thirty years show larger and larger increases of insanity, which can in no other way be reconciled with rational views on the subject than by assuming each census, compared with its successor, to have been wanting in comprehensiveness.

There is another important point to be considered in this connection. Even from a superficial comparison of the different censuses, it must become patent to the average intelligence, that with them, as in all human pursuits, practice makes perfect. From decade to decade the censuses have grown more valuable, as well in the number of subjects investigated, as in the minuteness of detail and the exactness of classification.

In no respect are these improvements more obvious than in the subject under consideration. Thus, in the census of 1870, the insane population is classified only by race, place of birth, age and sex; while in that of 1880, it is also arranged according to places where found, separating insane in asylums and other institutions from those treated at home. This improvement over the method of the former year helps in some measure to explain the origin of the apparent increase, since it must be inferred that the work was done more thoroughly in 1880 than in 1870. If the census of the latter year had contained a classification of insane according to "places where found," it would probably be easy to demonstrate that the increase of insanity in excess of the ratio of growth in population, must be placed under the head of "insane at home"; as it is self-evident that the full enumeration of this class must present greater difficulties than that of the insane in institutions.

The inevitable lack of accuracy in the census of 1870, in this respect, becomes still more obvious from the following comparative table, showing the proportion of insane to the entire population in a number of European countries and the United States:

ENGLAND:	Proportion	of insane	to population	......2.47	in one	thousand.
SCOTLAND:	"	"	"	.....2.96	"	"
IRELAND:	"	"	"	.....3.30	"	"
FRANCE:	"	"	"	.....1.33	"	"
BELGIUM:	"	"	"	.....1.49	"	"
SWEDEN:	"	"	"	.....1.94	"	"
DENMARK:	"	"	"	.....1.97	"	"
UNITED STATES:	"	"	"	.....0.97	"	"

Although the difference between the condition, mode of life, temperament and habits of the laboring population of some of these European countries, and those of residents in the United States would, without further comment, explain to a certain degree the above contrasts, it would nevertheless be almost quixotic to ascribe the whole difference, flattering as it may be to our national pride, to these conditions exclusively, rather than, partly at least, to defective enumeration by the census takers here.

There is no doubt an increase of insanity, exceeding the ratio of increase in population; but, unless extraordinary agencies can be shown to have essentially changed the condition of the people within the decade, there seems to be no reasonable ground for the assumption that this increase amounts to one hundred per cent., as shown by the census. Eminent physicians, with whom the compiler of this sketch had the honor of conversing on this subject recently, are of the opinion, for the reasons stated, that the reported increase is to a certain extent apparent only.

EVIDENCES OF DECREASING INEBRIETY.

Let us grant, however, for the purpose of argument, that the increase is real, that is to say, that there were in the United States, in 1870, no more insane than the census of that year accounts for, and that the proportion of insane to population actually rose from about one tenth of one per cent., or one in one thousand, in 1870, to one-fifth of one per cent., or two in one thousand, in 1880; from which it would appear that, compared with the growth of the population, there is an abnormal increase of insanity amounting to nearly one hundred per cent.

If we furthermore assume that this enormous increase is due, as advocates of prohibition claim, to increasing inebriety,—which, by the way, could not take place in any civilized community without being attended by many other palpable phenomena of a general demoralization—what could be more natural and reasonable than to conclude that direct statistical evidence sustaining such assumptions would be sought to be obtained by those who most need it?

No attempt was ever made, nor is any likely to be made, by the opponents of moderate laws to conclusively demonstrate by statistics of causes of insanity the correctness of their views; and this appears all the more remarkable, when we remember the fact that large sums of money are annually expended by the Temperance Publication House for books, which, as propagators of temperance ideas, cannot begin to be compared with what such a statistical exhibit would prove to be, if the assumptions of ultra-pessimists were correct. Nor will the impartial reader be apt to think that this indifference to so potent an auxiliary springs from any desire to deal leniently with those misguided creatures, who believe that it is the law's province to discriminate between drinker and drunkard, instead of tyrannizing over the former in order to reform the latter. Whether they have a sovereign contempt for the details of statistics, charity for their opponents, an apprehension of forging weapons for their adversaries—or whatever else the reason, the fact remains that prohibitionists have not only failed to collect such new data as seem necessary to sustain their assertion, but they have even ignored those already collected facts, which may well serve as a trusty guide to honest truth-seekers.

The necessity for the present inquiry will, in view of these circumstances, be readily understood. Some one had to undertake the task of establishing, so far as practicable, a statistical basis for all future discussions on this subject, and thereby to restrain the exercise of highly fertile imaginations within the bounds of at least a semblance of reality.

Even without such a basis it would, however, be no difficult task to reduce the position of prohibitionists *ad absurdum*. For, while no sane man will venture to gainsay that intemperance is a source of insanity, yet so numerous and obvious are the indications of a decrease of inebriety, that it seems preposterous to assume that the ravages of this vice have increased in the ratio represented, or anything remotely like it. To show that none are more blind than those who do not wish to see, it may not be amiss, before proceeding to a consideration of our statistics, to briefly review the indications that warrant an assumption diametrically opposite to that of the prohibitionists.

Increase of inebriety not only means augmented consumption of intoxicating drinks generally, but an increased consumption of a certain kind and quality of drinks, and that is not all; it also implies impairment of the moral sense and a general deterioration of either the social, political or material condition of the people. Wars, protracted depressions of business, industrial crises, famine, epidemics, political oppression and like ills are not generally preceded, but almost invariably succeeded by a laxity of morals, one single feature of which may be, and usually is, intemperance. All these things must be considered, if, in the absence of statistics, a correct judgment is desired. For purposes of a rough estimate it may suffice to take quantity and kind of drinks consumed as a basis in determining the extent of inebriety; but a more exact conclusion will be reached by carefully weighing all correlative conditions, not forgetting that the physical and mental effects of the diverse kinds of drinks vary essentially in different climates. Indeed, on this climatic influence Montesquieu bases his distinction between national and individual inebriety. In his "Spirit of the Laws," he says: "Pass from the equator to our pole and you will find inebriety augmented with the degrees of latitude." Dr. Bowditch, one of the most learned and ardent advocates of vine-culture in this country, formulates a similar conclusion, when he says: "Intemperance prevails the world over, but it is very rare at the equator. The tendency increases according to latitude, becoming more brutal and more disastrous in its effects on man and society as we approach the northern regions."* Montesquieu maintains that in countries "where the vine is indigenous,

* "Report on the Use and Abuse of Intoxicating Drinks throughout the Globe," Mass. State Board of Health, 1872.

inebriety has few evil effects on society,"* and it is a well-known fact that in such countries the use of beer and ale may be liberally indulged without great detriment to the physical and mental well-being of the drinker.

Now, in our case it can be shown:

1. That the consumption of distilled spirits has decreased during the decade in question, and that a decline in the use of ardent spirits has steadily been going on since 1820.

2. That the consumption of fermented beverages, best suited to the people of our country (the Vineland of the Vikings) has become greater and more general, essentially changing the drinking habits of the people.

3. That the condition of the people has been ameliorated to an uncommon degree.

Consumption of Distilled and Fermented Drinks.—In 1870 the aggregate quantities of all kinds of distilled spirits upon which the United States revenue tax was paid, amounted to 78,490,198 gallons, of which, according to the unreliable mode of calculation generally adopted, one-third was used for manufacturing purposes, leaving about fifty-two million gallons for consumption as drink by a population of 38,115,641 souls. In 1880 the aggregate quantities of all kinds of distilled spirits, upon which the tax was paid, amounted to 62,132,415 gallons, of which, if again only one-third be deducted as the quantity used for manufacturing purposes, about 40,000,000 gallons remained for consumption as drink by a population of over 50,000,000. According to this computation, the per capita consumption of domestic distilled spirits was about 5 quarts in 1870, and 3⅕ quarts in 1880. In the former year, the excess of imports over exports amounted to 615,560; in the latter year we imported 1,606,084 gallons, and exported 11,504,741 gallons. It is assumed, upon what authority we know not, that the per capita consumption of distilled spirits in 1880, was 4⅕ quarts, taking this to be the correct figure, we would in view of the difference in the quantities for which tax was paid in those years, have to infer that the per capita consumption in 1870, was 6 quarts.

* "Il est naturel que, là oû le vin est contraire au climat, et par conséqnent à la santé, l'excès en soit plus sévèrement puni que dans les pays ou *l'ivrognerie a peu de mauvais effets pour la personne,* oûelle en a peu pour la société," etc.

With a few intermissions of an upward tendency, caused by changes in the tax rate, the decrease in the consumption of distilled spirits has steadily gone on for a great number of years, keeping pace with the growth and development of the brewing industry.

In 1817, the consumption of distilled liquors in the United States amounted, according to Bristed, to 25,000,000 gallons. The population at the end of that decade (1820) was a little over nine and a half millions, hence the per capita consumption, roughly estimated, amounted to over 11 quarts. At that time there were about 15,000 distilleries in operation, but only 113 breweries. By a large import duty imposed on beer, ale and porter, the growth of the brewing industry was retarded rather than accelerated, as was obviously intended; for, although a like duty was imposed on imported distilled spirits, while the home-manufacture was subjected to but trifling restrictions, yet such were the habits, mode of life and tastes of a greater part of the people, that they naturally gave preference to ardent spirits. Here then, comparing 1820 with 1880, we find a decrease from 11 quarts in the former year to 4⅕ quarts in the latter. As to the quality of distilled drinks which, as has been said, is an important factor in causing insanity, the past generation appears to have been no better off than we of to-day, if any credence can be placed in the words of Morewood and other writers of that period.*

The drinking habits have necessarily undergone an essential change. Solitary drinking, which at one time must have prevailed to a considerable extent, has almost entirely gone out of fashion.† In its stead we have, it is true, the evil of treating at the bar; but this perpendicular system, as Dickens styled it, is growing less at the same rate as the taste for fermented drinks, with their accom-

* In his "History of Inebriating Drinks," that author says: "The distilleries (of the United States) for the most part are conducted on small scales; and, as might be expected when the trade is committed to a vast number of people of opposite interests, a great deal of competition as well as ignorance prevails. Breweries not being generally established, the want of barm has not failed to produce great inconvenience, and the distillers are obliged to have recourse to *deleterious substitutes* for the fermentation of their wash. Hence arises that ardent quality which renders their whiskey in many instances disagreeable to foreigners."

† In the United States, the grand source of temperance reform, it was, previous to the introduction of temperance societies, considered as nothing shameful for men to drink liquor by themselves. Indeed, at that period, solitary drinking was there an admitted practice. Here (in England) so strong is the general feeling on the subject, that many open drunkards would abhor the idea of being convicted of solitary drinking.—*J. Dunlap's "Drinking Usages of the United Kingdom."*

paniment of sociability, music and innocent merriment, becomes more general.

The production of fermented beverages increased from 6,574,617 barrels in 1870, to 13,347,111 barrels in 1880. The influx of Germans accounts, of course, for a great part of the extraordinarilyrapid growth of the brewing industry; but at least one-third of it must be placed to the account of a radical change in the taste and drinking habits of a large proportion of the native population.

In order to fully appreciate the effects of this salutary change, particularly in connection with the question mooted, it should be borne in mind that in all countries where fermented beverages, such as light wine, beer and ale are the every-day drink of the people, alcoholism and its results are comparatively rare—an accepted fact that has prompted wise governments of countries, whose people are addicted to the excessive use of distilled spirits, to enact laws the tendency of which is to encourage the use of fermented drinks. Unwise legislation has frequently been resorted to for the forcible suppression of the manufacture and use of ardent drinks, but has signally failed in every instance. Not to prohibit the use of the latter beverages, but to create a taste for lighter drinks, seems to be, judging from numerous experiences, the essence of efficacious laws on the subject.

Thus we find that the temperance agitation in the Netherlands, where the people suffered more from the abominable quality of the poorest and cheapest kind of gin than from the excessive quantities consumed, culminated in a law which places the utmost rational restrictions upon the sale of gin, while it encourages the use of beer. And the operation of this law, enforced since 1881 throughout the land with great severity, is officially announced to have already resulted in an appreciable decrease of mental derangements resulting from intemperance. Intemperance is really not a proper term in this connection, since even the moderate use of such execrable stuff as is being sold to the laboring people of the Netherlands would necessarilly affect the brain. It will be shown presently, that the change from bad to *good* whiskey and brandy has also effected the objects of temperance advocates in Sweden.

The working of the new liquor law has been the subject of much interest, mingled with grave apprehensions and anxious solicitude, not only in the Netherlands, but throughout Europe; the phases of its development have been closely watched and investigated by government officials, and the showing, so far, is most favorable.

Dr. van Cappelle, an officer of the Dutch Department of the Interior, recently published the following table.

	Number of Persons Admitted to Asylums.	Abuse of Alcoholic Drinks Assigned as Cause.	Number of Cases of Acute Mania.	Number of Cases of Acute Mania Caused by Intemperance.
1878	564	96 (17%)	132	34 (25%)
1879	592	119 (20%)	115	46 (40%)
1880	563	84 (15%)	125	32 (25%)
1881	580	92 (15%)	139	37 (26%)
1882	573	72 (12%)	125	31 (25%)

The per capita consumption of gin has decreased from 9.81 litres in 1881, to 9.46 litres in 1882. Undoubtedly other measures, simultaneously adopted, and directed against deep-rooted evils in the condition of the laboring classes, have materially contributed to this result, yet all the official reports, as well as the movements of the temperance societies—upon whom the government relies to a great extent for the realization of its reformatory plans—indicate that the eradication of the evils complained of is thought to be dependent upon the degree of success which will attend the effort to supplant gin by beer. The temperance societies of the Netherlands proceed in a very practical manner in this matter, as will be seen from the following excerpt from the report of a German Committee of Inquiry, sent to Holland to ascertain the results of the new law:

The temperance associations were actively and successfully engaged in the struggle against the abuse of distilled liquors. In the report recommending the use of beer as the most wholesome beverage for the people, the Commission move that they, on behalf of the Association, be intrusted with the supply of such kinds of beer as answer the demands of the Commission in every respect. They recommended nine breweries, the beers of which were found to be excellent, containing no more than 4% of alcohol, and from 5% to 8% of malt; five other breweries were recommended, the beers of which, excellent as they are, contain only from 3% to 5% of malt, and eleven breweries, where beers are brewed which contain 3% of alcohol and from 3% to 5% of malt. The report moves, further, that a commission of three members be appointed to control the brewers, to examine the beer several times a year in order to find out whether it always answers the demands, to discontinue taking part of the supply from those breweries where the beer is found not to answer the purpose any longer, and to give orders to other breweries who will comply with the wishes of the Association The delegate from Haarlem regretted that the otherwise excellent report did not demand the abolition of the beer tax, which he called inadmissible and immoral. A member of the Commission did not deem it wise to demand the abolition of the beer tax, the income from which, he said, aggregated to about 800,000fl., and which did not increase the price of the beer ; the abolition, therefore, would not involve any practical improvement. The delegate of the Amsterdam District Association read a communication in which particulars were given of the results of the three beer saloons established by the Association and in which one third of a litre of good, light beer was sold for five cents. The financial result was a success, as on some days 5,000 glasses of beer had been sold. In one of the halls the German delegates tried the beer and found it to be very good.

Here we have temperance societies—surely no less sincere and ardent, even if they be a little more rational and genuinely humane in their efforts than our good prohibitionists—striving to foster an industry which General Dow would fain root out completely.

In the capital of Bavaria inebriety is very rare, and alcoholism, of course, still more so. Only the very lowest persons—incorrigibly dissolute characters—drink ardent spirits there. The following letter needs no comment in this connection :

MUNICH, February 28, 1884.

G. Thomann, Esq.

DEAR SIR :

Alcoholismus chronicus is very rare in our hospital, because our laboring people drink beer. Distilled liquors are used very seldom.

We have had in the last year (1883) in our hospital nearly ten thousand patients, but only twenty-one cases of alcoholism.

I have the honor, dear sir, to be yours,

DR. VON ZIEMSSEN,
Professor, and Director of the General Hospital of Munich.

In the south of France, where wine is the common beverage of the people, alcoholism, according to a scientific report of recent date, is of much less frequent occurrence than in the north, where distilled spirits are ordinarily used. French reformers have always deprecated the use of ardent spirits because of many adulterations, recommending the use of light wines as best suited to the temperament of the people. Professor A. Bouchardat, formerly of the "Medical Faculty of Paris," though "vigernon de naissance et de cœur"(wine-grower by birth and choice), as he styles himself, went still a step further. In discussing the dangers of the excessive use of distilled spirits,* he writes enviously of the quantities of beer consumed in Germany and England, and gives utterance to the hope that the use of beer may yet become more general in France. "Good beer," he says, "is the most wholesome of fermented beverages. Its continued use from the remotest ages to our day bears sufficient testimony to its excellent qualities." What a world of significance in these words, coming as they do, from a *savant*, whose country's wealth is in no small measure represented by its vine-clad hills! Since they were written the excellent qualities of beer have conquered French prejudices against a beverage so thoroughly German.

History presents many illustrations of the reformatory uses to which beer has been applied with success; and also many proofs of the pernicious results of legal and natural restrictions placed upon the manufacture and use of fermented drinks. Smollett, whom some regard as the precursor of the temperance advocates of our day, and whose writings we find frequently cited in the Rev. William Reid's "Temperance Cyclopædia," was of the opinion that "the best way of preventing the excess of spirituous liquors would be to lower the excise on beer and ale, so as to enable the poorer class of laborers to refresh themselves with a comfortable liquor."† The exceedingly high malt tax had driven the English people to the use of gin, and with what effect may be seen from Smollett's graphic chapter on high licenses. He says: "When those severe duties (amounting almost to prohibition) were imposed, the populace of London were sunk into the most brutal degeneracy by drinking to excess the pernicious spirit called gin, which was sold so cheap that the lowest class could afford to indulge themselves in one continued state of intoxication." And here, again, it was not so much the quantity as the bad quality of the drink consumed that caused the evil effects. High licenses, as will be shown in another part of this sketch, made things rather worse than better; one unwise measure being succeeded by a still more foolish law.

Prohibitionists who appeared before a committee of the Massachusetts legislature, some weeks ago, asserted that the present English Beer Bill did not diminish drunkenness. Well, that is surely not the fault of beer, any more than it was at the time of the famous Gin Act that Smollett wrote about.

* L'eau de vie, ses dangers.

† History of England; chap. XXXII.

If people will not drink beer, having been educated by unwise legislation to the use of ardent spirits, they cannot be held up as examples of what fermented beverages will do for body and mind.

It is true that at the time when beer and wine were yet the common beverages of the people of England and Germany, inebriety was no rare thing in either country, and exhortations against intemperance from pulpit and rostrum were frequent. But it must not be lost sight of, that these exhortations were directed against intemperance in much the same spirit in which rebuke was administered for gluttony, passion for dress, or like excesses. Drinking to excess has been regarded as a national evil in England as well as in Germany, from the very earliest ages. When the Romans invaded Britain they found the vice in full growth, and it is well known what horrible stories Tacitus tells of the drinking habits of the Germans of his time. But one will search in vain for the slightest shadow of evidence in history that the knowledge of this evil was linked with a fear of a degeneracy of race. Indeed, if these excesses, uninterruptedly continued in both countries from the time of the Roman invasions, had a degenerating influence on the people, the histories of the two nations fail to reveal any indication of it. Even early restrictive laws directed against intemperance are by no means evidences of threatening moral and physical degeneracy of race. For when, on the advice of Dunstan, Archbishop of Canterbury, Edgar, in 958, issued his edict against alehouses—an act frequently cited as an early precedent for restrictive measures—it could still be said of the English people, "that in the midst of all vicious and sensual indulgence the clergy of that day trod the path of sanctity, and that many of the laity, of all ranks and conditions, were well pleasing to God." And we can surmise what the physical condition of that nation must have been, which in the nine succeeding centuries has erected a vast empire and swept the armadas of the eastern world from the seas.

In Germany the intemperate use of beer and wine had reached its height in the sixteenth century, sometimes styled "the century of drinking bouts;" and, indeed, the chronicles of some of the performances at the convivial boards of that time appear almost incredible. Yet after the Thirty Years' War had devastated the vineyards in the south of Germany, and distilled spirits of a most abominable quality had been introduced by the martial hordes from the North, the debauches of past periods seemed like harmless pastimes. The "horrors of alcoholism" then began for the first time to be known on the banks of the Elbe, as well as on those of the Rhine, although as yet no uniform technical term had been found for the evil.

Want, suffering, despair, and laxity of morals aggravated the sad consequences; but the most destructive factor was again the bad quality of the drink distilled from potatoes.

A negative illustration of the beneficent influence which beer exercises on the moral and mental condition of men, is afforded by the present state of things in Baden, a report on which is published in the organ of the German temperance societies, *The Nordwest*, of February 18th, 1884. It reads:

The grand duchy of Baden is, generally speaking, not a soil on which the weeds of alcoholism will thrive. A good, light wine is raised almost throughout the whole country, and thus the peasantry are protected against the use of strong liquors; besides, the large manufacture of beer, facilitated by an abundance of grain and hops, grown in that country, removes pretty well the danger of the use of liquors on the part of the working classes.

An exception to these natural conditions is found in the Black Forest, where the sharp air, much colder weather, outdoor work, felling of trees, blasting of rocks, plowing, etc., have driven the population to the use of alcohol. The property of the peasant of the Black Forest consists in cattle, meadows and fields; what he owns besides these things is not much. The meadows belong to the cattle, while the fields provide for the wants of men; and on these fields the peasant will grow summer wheat and potatoes alternately; three-fourths of his food are potatoes and one-fourth consists of bread, milk and pork. This coarse food, which at the same time is hard to digest, together with hard labor, induces the farmer to use distilled liquors, the more so as country-wine, beer or cider is not easily to be obtained. Now, it has been generally observed that during recent years the use of liquors has become a habit with the inhabitants of the plains, but only because there has not been a really good wine-year since 1874 and 1876; in some parts of the country the wine crop has even been an utter failure. But nobody doubts that, as soon as the wine crop improves, the evil will disappear; and in order to put a stop to the habitual use of brandy, which in the Black Forest amounts to one-quarter litre daily, per capita (in 1881 it was one-ninth litre per capita in the "Amt" of Mühlheim), the Minister of the Interior has especially directed the authorities of the different circuits to combat this evil by all legal means. Thus, we see, that our government has a watchful eye on the use of ardent liquors and the sale of the same. But experience teaches us that such laws cannot be strictly enforced. In the rural districts the use of liquors will certainly decrease. It may even disappear altogether, as soon as some good wine and fruit crops enable the poorer people to obtain a cheap and wholesome homemade beverage. It will prove an impossibility to abolish this evil entirely, so far as the Black Forest is concerned; but it can be lessened by the sale of some light beer and by the use of homemade beverages. What the government has attempted in this direction has not met with success."

Thus in Baden, as in most all civilized countries, the government endeavors, by all means within its power, to accomplish what the good sense of the American people, aided, no doubt, by the noble efforts of rational and honest temperance advocates, has achieved in a great measure, and continues still to strive for.

Since Alsace-Lorraine has become part of Germany, the price of wine, formerly the ordinary drink of the people, has risen to such a figure, that the poorer classes can no longer afford to use it. They now drink ardent spirits, distilled from prunes or cereals and the result is, that alcoholism becomes more frequent. The director of the insane asylum at Brumath, Dr. Stark, reported, in 1880, after an investigation covering a period of six years, that 29 per cent. of his male patients had been inebriates previous to their admission. The climate as well as the habits of the people require a milder drink there.

From a comparative table, based upon authentic statistics, it appears that in 1873 the rate of mortality from alcoholism was highest in Stockholm, Sweden; lowest, except one other place, in the three Bavarian cities: Munich, Würzburg and Nürnberg. In Stockholm the rate among men was 1.95, among women 0.12; while in Munich the rate among men was 0.15, and naught among women. In the latter city, the capital of the land of beer-drinkers, the per capita consumption of beer in that year was 250 maas, or about 62 gallons. In the duchy of Nassau the consumption of distilled spirits had assumed uncommonly large proportions up to the year 1840, and in consequence alcoholism had became so frequent, that the government felt constrained to adopt measures furthering the object of temperance societies in introducing good beer for general consumption. Since then mental derangements from that cause have steadily grown less.

Such evidence could easily be made interminately cumulative. Enough has been said, however, it is hoped, to show that a decrease in the consumption of distilled spirits, attended by an increase in the use of fermented beverages, in countries were the vine is indigenous, must necessarily diminish that form of intemperance, from which, when persisted in to excess, alcoholism and insanity are apt to result.

Is there any reason for excepting our country from this universally acknowledged and oft-tested rule? And if there is none, how can it be claimed, in the face of so many conditions favoring a decrease of inebriety, that insanity has increased at such an enormous rate, because of the intemperate habits of the people?

Anyone wishing to see cannot fail to perceive, that from 1870 to 1880 inebriety must have been less operative as a cause of insanity than in former decades. If prohibitionists cannot see this, it must be supposed they do not desire to see it, probably wishing for the contrary. This is no idle supposition. Colonel Lucius B. Marsh, who recently appeared before the Police Commissioners of Boston to urge the high license system on behalf of the prohibitionists, said that his party wants the evils of the liquor traffic to become more and more grievous, so that their remedy shall be adopted perforce in the end.* Can an impartial and unbiased judgment be expected from such sources?

Condition of the People.—Notwithstanding the business depressions which during part of the decade, from 1870 to 1880, weighed heavily upon some industries, the general prosperity of the workers has enormously increased, the material progress of the whole people nigh bordering on the fabulous. Statisticians agree, that from the beginning of time no such progress was ever recorded for so brief a period, as that made by the United States from 1870 to 1880.

A more striking picture of our material development cannot be conceived than that outlined in Mulhall's Balance-Sheet of the United States, the essential features of which may thus be summed up: Rise of aggregate American industries, 35 per cent.; actual increase of industry 525 millions, exceeding the maximum among European nations (Great Britain) by 188 millions; excess of exports over imports, 31 per cent.; rise of manufacture, 30 per cent.; rise of mining, 90 per cent.; increase of farming stock, 33 per cent.; increase of carrying trade, 23 per cent; increase in mileage of railways and telegraphs, 100 per cent.; increase of net income per inhabitant, 5½ per cent.; decline of bankruptcies, 50 per cent.; reduction of national debt, 22 per cent.; excess of grain supply over consumption 18½; excess of meat-supply over consumption, 36 per cent.; 30 per cent. of the grain and 30 per cent. of the meat of the world being produced by the United States.

Europe with a population of 327 millions produced 5,272 million bushels of grain, and consumed 5,652 million; the United States, with a population of 50 millions produced 2,390 million bushels of grain and consumed 2,020 million, from which Mulhall draws the conclusion that the "Americans are apparently the best fed of all nations." Surely, in all this there is no ground on which to base the assumption that drunkenness,—quite as frequently caused

* N. Y. Sun, February, 1884.

by want as by anything else—has increased a hundred per cent., or that it has increased at all.

Those who "guess" that intemperance is the cause of insanity in fifty or in twenty-five, or in twenty, or even in fifteen cases out of a hundred, might, if we had nothing better to offer them, form an approximately correct opinion from the following authentic table:

Name of Country.	Per capita consumption of Distilled Spirits.	Proportion of Intemperance to other causes of insanity.
Denmark *..............	18 quarts.	19 %
Sweden †................	11 "	14 %
North Germany..........	10 "	13 %
Holland..................	9 "	12 %
United States............	4 1/5 "	?

The interrogation point in the second column opposite the name of our country will presently be replaced by a figure.

INTEMPERANCE AS A CAUSE OF INSANITY STATISTICALLY CONSIDERED.

The question whether intemperance has been more active as a cause of insanity during the ten years, ended in 1880, than in the preceding decade, can definitely be solved by comparative statistics only; unless, indeed, both inductive and deductive reasoning be resorted to, in which case, as we have seen, the question would have to be negatived.

Underrating the difficulties to be overcome, the compiler originally intended to prepare comparative statistics of insanity caused by intemperance, and likewise of all forms of alcoholism. But at the very beginning, this comprehensive plan had to be abandoned for one whose execution seemed less likely to meet with insuperable obstacles, so that ultimately the inquiry was confined to institutions for the treatment of the insane, instead of being extended to hospitals also.

But even so, the investigation which in spite of many discouraging impediments, has been pursued with considerable persistence for six months, failed to yield one of the essential points of the result sought for, viz.: the means for a comprehensive comparison with former years. Indeed, to be plain with the reader, no amount of labor, however judiciously applied, will ever wholly sup-

*† These figures and all other information contained in this sketch, and relating to Denmark, Sweden and Norway, are taken from official reports and statistics, for which the compiler is indebted to the kindness of Mr. Christian Boers, Swedish Consul, and Mr. T. Schmidt, Danish Consul, both of New York. Information relating to other European countries has been obtained from original sources directly.

ply this lack. The fact is, that in very many institutions the records have been kept in such a manner as to preclude the possibility of ascertaining the causes of insanity in cases treated during past years; while in a great many other asylums the mode of inquiry into the history of patients was such, that no one but the inquirer himself could, with any degree of certainty, determine real or assigned cause. But while the information contained in Appendix A is not altogether satisfactory as a basis of general comparison, it affords ample material for an entirely reliable judgment as to the present extent of the relation which intemperance bears to insanity. Appendix A comprises statistics from fifty-four institutions, in addition to letters from the superintendents of twenty-six more. In all, eighty institutions are accounted for in some way or other; in this number are included nearly all the larger institutions in the country, and the geographical groups of States are pretty equally represented.

According to the census of 1880 there were, in that year, one hundred and forty-nine hospitals for the treatment of the insane; forty of these are very small institutions, having from two to ninety patients; their aggregate number of patients being only 1261, out of a grand total of 91,997. Inasmuch as very few of these smaller asylums have been heard from, it may safely be asserted that three-fourths of all the larger institutions in the country have contributed to Appendix A.

In view of the fact that four successive circular letters have been addressed to all those superintendents of insane asylums, who failed to reply, it might not unreasonably be assumed that these physicians, doubting the legitimacy or justifiableness of such an inquiry, are unwilling to give the necessary information, or that they are unable to do so from a want of explicit records. This, it is deemed expedient to state distinctly, in order to avoid a possible reproach of unfair discrimination or biased dealing. Throughout his self-imposed task the compiler has been guided by no other motive or desire but that of presenting to the public the whole truth; neither more nor less. It is for this reason that he has in no instance changed, or essentially abridged the letters of transmittal accompanying statistics, or the letters containing only opinions. The benefit of doubt has consistently been bestowed on the opposing side; as, for instance, in the case of the report from the Milwaukee Asylum, in which, as appears from the heading, are included cases "classified as inebriates." * This occurs in a few more instances, and tends, of course, to swell the average ratio. But no deduction was attempted by the compiler in any instance.

Without going "behind the returns" in any case; without arbitrarily changing or correcting any figures, but giving them and accounting for them strictly in accordance with the intentions of the physicians from whom they emanate, and without availing himself of the benefits of any dubious information—even where better judgment seemed to dictate such a course—the compiler is bound to conclude from the facts before him that, on an average, intemperance is the cause of insanity *in seven cases out of one hundred.*

Following are the figures upon which this conclusion is based:

NAME OF ASYLUM.	NUMBER OF INSANE.**	NUMBER OF CASES OF INSANITY CAUSED BY INTEMPERANCE.			PERCENTAGE OF INSANITY CAUSED BY INTEMPERANCE.
		Male.	Female.	Total.	
1. Alabama Insane Hospital, Tuscaloosa	210	19	...	19	9.048
2. State Asylum, Napa, California	563	31	5	36	6.358
3. " Stockton, "	106	8	....	8	7.547
4. Lunatic Asylum, Pueblo, Cal.	† 100	10	....	10	10.000
5. Hospital for the Insane, Middletown, Conn	† 100	20	....	20	20.000
6. Retreat for Insane, Hartford, Conn.	78	4	1	5	6.416
7. Govt. Hosp. for the Insane, Washington, D. C.	265	54	1	55	20.774
8. Illinois Central Hosp. for the Insane, Jacksonville, Ills	215	8	....	8	3.720
9. Southern Hospital for the Insane, Anna, Ills	157	1	....	1	.634
10. Eastern Hospital for the Insane, Kankakee, Ills	188	17	2	19	10.106
11. Indiana Hospital for Insane, Indianapolis, Ind	698	13	2	15	2.149
12. Iowa Hospital for the Insane, Mount Pleasant, Iowa	† 100	25	....	25	25.00
13. Iowa Hospital for the Insane, Independence, Iowa	871	18	1	19	2.181
14. Western Kentucky Lunatic Asylum, Hopkinsville, Ky	100	7½	....	7½	7.50
15. Mt. Hope Retreat, Maryland	585	36	18	54	9.230
16. McLean Asylum, Somerville, Mass.	237	9	....	9	3.797
17. Boston Lunatic Asylum, Mass.	104	10	....	10	9.615
18. Taunton " "	786	25	5	30	3.817
19. Danvers " "	721	64	24	88	12.205
20. Eastern Michigan Asylum, Pontiac	806	21	6	27	3.349
21. Michigan Asylum for the Insane, Kalamazoo	¶ 3,534	159	....	159	4.499
22. Minnesota Hospital for the Insane, St. Peter, Minn	253	26	....	26	10.276
23. Mississippi State Lunatic Asylum	429	1	....	1	0.233

* See number 80 of Appendix A.

NAME OF ASYLUM.	NUMBER OF INSANE.**	NUMBER OF CASES OF INSANITY CAUSED BY INTEMPERANCE.			PERCENTAGE OF INSANITY CAUSED BY INTEMPERANCE.
		Male.	Female.	Total.	
24. State Lunatic Asylum, Fulton, Mo	¶ 2,000	77	....	77	3.550
25. St. Louis Insane Asylum, Missouri	¶ 2,204	169	41	210	9.528
26. Missouri Lunatic Asylum, No. 2..	316	11	....	11	3.480
27. New Jersey State Lunatic Asylum, Trenton, N. J....	633	12	....	12	1.895
28. Erie County Asylum, Buffalo, N.Y.	339	2	1	3	.885
29. Hudson River State Hosp., Poughkeepsie, N. Y................	277	22	2	24	8.664
30. State Lunatic Asylum, Utica, N.Y.	404	36	2	38	9.406
31. St. Vincent Refuge for the Insane, Harrison, N. Y.........	39	....	1	1	2.564
32. State Homeopathic Asylum, Middletown, N. Y............ .	509	5	....	5	0.982
33. Marshall Infirmary, Troy, N. Y...	274	109	6	115	41.971
34. Buffalo State Asylum, Buffalo, N.Y.	265	28	7	35	13 207
35. Sanford Hall, Flushing, N. Y...	45	2	1	3	6.666
36. Dayton (Ohio) Asylum............	¶ 5,409	221	10	231	4.270
37. Columbus (Ohio) Asylum.........	290	16	...	16	5.517
38. Longview Asylum, Carthage, Ohio	¶ 4,647	398	85	483	10.396
39. Athens (O), Asylum for the Insane	216	13	3	16	7.407
40. Cleveland (Ohio) Asylum	244	18	....	18	7.377
41. State Hospital for the Insane, Norristown, Pa.......	532	31	....	31	5.827
42. State Hospital for the Insane, Warren, Pa.................	423	6	1	7	1.654
43. Lunatic Hospital, Harrisburg, Pa..	64	4	3	7	10.937
44. Asylum for the Relief of Persons Deprived of the Use of their Reason, Philadelphia, Pa.......	¶ 1,063	110	8	118	11.100
45. Pennsylvania Hospital for the Insane (Kirkbride), Phila., Pa....	¶ 2,212	285	....	285	12.884
46. Butler Hospital, Providence, R. I..	195	7	4	11	5.641
47. South Carolina Lunatic Asylum, Columbia...	603	19	2	21	3.482
48. Tennessee Hospital for the Insane, Nashville.....................	408	5	2	7	1.715
49. Vermont Hospital for the Insane, Brattleboro, Vt..............	100	$6\frac{2}{5}$	....	$6\frac{2}{5}$	6.40
50. Western Lunatic Asylum, Staunton, Va......................	534	32	2	34	6 386
51. Eastern Lunatic Asylum, Williamsburg, Va	536	9	....	9	1.679
52. Central Lunatic Asy., Richmond, Va.	184	15	8	23	12.500
53. West Virginia Hospital for the Insane, Weston, W. Va.........	685	59	....	59	8.613
54. Milwaukee Asylum for the Insane, Wauwatosa, Wis..............	117	21	....	21	17.946
	36,973	2,334	254	2,588	......

Total average percentage, 6.99.

** The numbers given in this column are those of patients admitted either during the year 1883, or the most recent year for which statements could be obtained. Figures marked † are assumed, the ratio only being given by the superintendents. Figures marked ¶ represent the aggregate number of patients admitted during a number of years.

Statistics from the asylum on Ward's Island, the largest in the country, could not be included in the table, having been received too late for that purpose. The figures, to which attention is particularly invited, will be found at the end of Appendix A. They could in no manner have changed the general result, inasmuch as Dr. Macdonald, the superintendent of the asylum, writes that only a relatively small number of those given under the heading "Intemperance" is *solely* due to that cause, but that in the majority several causes coöperate." Such cases could not properly have been included in our table.

Undoubtedly, the table is not wholly correct; but prohibitionists should not ignore the fact that it is the very best statistical exhibit that can be prepared with the attainable material; that no other material can be obtained from official sources, and that all its defects tend to render the ratio larger than it actually is.

There is an error in the table that could not be corrected, as the matter had already been electrotyped when it was discovered. The percentage at the Marshall Infirmary, Troy (41.97) seemed rather high, and an explanation was asked, under the impression that the Infirmary might be an institution for the cure of inebriates. The following letter was received in reply to our inquiry:

Medical Superintendent's Office,
Marshall Infirmary and Rensselaer County Lunatic Asylum,

TROY, N. Y., *March 21st, 1884.*

DEAR SIR: Your letter of the 5th was mis'aid, and has just now come before me.

The statistics I sent you relate to the sick department of this institution. If it is not too late, I will furnish those of the Asylum, if you desire them.

Yours respectfully,

JOS. D. LOMAX.

It is easy to see how much smaller the average ratio would have been without this error.

In examining the statistics and comparing it with the opinions expressed by superintendents of asylums, one will scarcely be able to suppress a degree of astonishment at the glaring disparity which exists between them in many cases.

Dr. H. A. Gilman, Superintendent of the Hospital for the Insane, at Mount Pleasant, Iowa, for example, writes: "I am prepared to say that about twenty-five per cent. of the cases of insanity, in something more than seven thousand patients that I have been familiar with, has resulted from the use of alcoholic beverages. . . . Twenty-five per cent. more may be traceable to same cause as a result of drunkenness in the parent." Dr. Gilman gives only a statistical estimate to bear this out; but an estimate, the correctness of which it would be unfair as well as impolite to doubt.

His colleague, Dr. G. H. Hill, Superintendent of the Insane Hospital at Independence, Iowa, furnishes detailed and positive statistics from 1874 to 1883; and from this it appears, that in the latter year 19 patients out of 871 were treated for insanity, caused by intemperance; the proportion being 2.19 per cent. Here, then, we have a difference of 22.81 per cent. between the statistical estimate of one physician and the statistics of another in one and the same State.

The difference here, as in other instances, may be explained in various ways; but for such explanations the reader must be referred to the letters contained in Appendix A. Much depends upon the disposition of the physicians to accept one or the other of the many theories bearing on this subject. This being a purely statistical sketch, no attempt can or will, however, be made to discuss this side of the subject. Yet it may not be out of place to quote the following from an article written by Dr. A. I. Thomas, of the Indiana Hospital for the Insane, and published in *The Indiana Medical Journal:*

Dr. Hammond, in his new book on insanity, classes alcohol as a potent cause of mental disturbance No one will dispute the fact that alcohol does properly bear a portion of the blame for insanity, but I do not consider it so puisant a factor as Dr. Hammond insists that it is. In a clinical experience of more than four years in an establishment for the treatment of insane persons my views upon this subject have been somewhat modified. At one time in my life I regarded alcohol as the cause of half of the cases of insanity, because I had been taught that such was the fact. Now I believe, of course, speaking from my experience alone, it produces a very small amount of such disease. My error was and is a common one, and is one into which both the profession and laity fall. It will always astonish one to examine the records of admission in the Indiana Hospital for the Insane, and note the small number of cases attributed to over-indulgence in alcoholic beverages. During the fiscal year ending October 31, 1882, there were 762 admissions—415 men and 347 women. In that number there were 26 cases—22 men and 4 women—rated as the victims of alcoholism, or, rather, indulgence in alcoholic liquors was given as the cause. A calculation will show that the percentage was a very small one, being a fraction over three and one-half per cent. of the total number.

Dr. Hammond and Drs. Bucknill and Tuke, of England, attribute to alcohol the production of a large number of cases of general paralysis of the insane. Dr. H. says twenty per cent. of his cases belong in that category. In one of the works of Bucknill and Tuke we find this observation: "Drink causing poverty, and poverty leading to drink (the former in by far the larger proportion of cases), are the familiar antecedents of an attack of general paralysis."

Dr. Mickle, of London, in his "General Paralysis of the Insane," says: "Alcoholic excesses are first in the list of causes."

Dr. Thomas J. Austin, of England, Medical Officer of the Bethnal House Asylum, differs with the authorities I have just quoted. In his book, which is the best treatise extant on this subject, Dr. A. gives the complete course of seventy-seven cases, as to cause, history, duration and post mortem results. In this number ten cases are attributed to intemperance in the use of alcoholic beverages. Dr. Austin says: "The ten cases which are attributed to intemperance will strike those who seem inclined to ascribe every ill that flesh is heir to to the abuse of alcoholic drinks as too few. However willing to admit drunkenness as a frequent source of physical disease, I very much doubt the truth of the reiterated assertion that it is often the immediate cause of insanity, and still more of general paralysis. It is more than propable that in the ten cases mentioned, the mind was already giving way when the incipient paralytic gave way to liquor, and that he flew to alcohol as a consoler to escape from that overwhelming care, in which is to be sought the true cause of his malady. That the disease is hastened by intemperance is likely; but inasmuch as the very characteristic physical symptoms, the result of intemperance are entirely wanting in general paralysis, drunkenness cannot be conceded as a primary, though in some cases, it is doubtless a powerful auxiliary cause."

For the purposes of this sketch it is not necessary to weigh the information, giving a high ratio, against that which shows a small proportion of insanity caused by intemperance. Neither the correctness of the one, nor the reliability of the other will be questioned; indeed, they cannot be questioned, unless the hope of ever establishing a statistical basis be at once relinquished.

The principal consideration is the average proportion, and this seems to be very nearly what the decrease in the consumption of distilled spirits would inevitably point out.

If a per capita consumption of distilled spirits of 18 quarts produces 19 per cent. of insanity in Denmark—a country where, according to the authorities already quoted, inebriety assumes a more brutal and disastrous form than here, why should a per capita consumption of 4½ quarts produce a greater proportion of insanity than seven per cent., in a country whose people enjoy a much greater degree of material prosperity than the Danes?

If the comparative table on page 23 be consulted, it will be found that even seven per cent. would appear to be proportionately too high.

A comparison with former years may be instituted on a somewhat general basis by adopting the statistics quoted by Dr. Baer, in his work on Alcoholism. It is stated there, reference being made to

Lunier's work on the same subject—that of 14,941 insane, treated in sixteen institutions in the United States, 1788 were deprived of the use of their reason by intemperance. The proportion is 11.97 per cent. Of 3,084 patients treated in the asylum at Worcester, Massachusetts, from 1833 to 1848, there were 322 whose mental derangement was caused by the excessive use of intoxicants. Dr. Edward Jarvis ascertained by direct inquiry that of 22,113 cases of insanity, 2,896 or 13 per cent. were traceable to excessive imbibition of alcoholic beverages. The latter inquiry must have been made about 17 years ago. Keeping in view the fact that the consumption of distilled spirits has decreased; that the drinking habits of the people have become more refined, and that the amelioration of the condition of the workers has removed a large number of incentives to inebriety, one can readily understand that the proportion of insanity should have decreased from 13 and 11 to 7 per cent.

The small proportion of insanity caused by intemperance in the Southern States finds its explanation in a number of well-known circumstances. The large colored population is very temperate and much less liable to alcoholism. The "Medical and Surgical History of the War of the Rebellion," furnishes striking proof of this assertion. Of the available force of white troops, averaging 431,237 men in the field and in garrisons, ten thousand came under medical treatment either for delirium tremens, inebriety or chronic alcoholism; of the available force of colored troops, averaging 60,854 men, only fifty required medical treatment for like causes. The proportion in the former case is 2.378 per cent.; in the latter 0.082.

In his annual report for 1883, the Surgeon General of the Army confirms these observations. On page 9, this point is elucidated in the following manner: "It is interesting to note that the colored troops make a particularly favorable showing in the small number of admissions for alcoholism and its results, exhibiting, as they do, a rate of only four (4) per thousand (1,000) to a rate of seventy six (76) per thousand (1,000) of mean strength among the whites. On the other hand, in diseases of the nervous system they have an unexplained preponderance."

The mild climate, imposing few hardships on the poor; the small urban population; the comparatively simple mode of life, and, probably, also the quality of ardent spirits consumed in the South, in a measure account for the small ratio of the results of alcoholism there.

The question of quality should not be underrated, seeing that it is almost as important a factor in the production of brain-diseases as the quantity consumed. The celebrated Gothenburg system (*) did not diminish the consumption of ardent spirits until very recently. The result of an investigation, conducted by a duly appointed Committee of Revision, shows that while from 1862 to 1866, the per capita consumption was from 3.85 to 4.37 *kans*, it rose to 4.15 *kans*, in 1872, and 5.14 *kans*, in 1876. Notwithstanding these facts—which, by the way, clearly demonstrate the utter futility of any attempt to regulate the appetites of the people, even by such a good measure as the Swedish system is said to be—the rate of insanity caused by alcohol has decreased appreciably, and so has the rate of mortality from the same cause. This decrease is, among other things, ascribed to the excellent quality of distilled spirits, the government having prescribed and maintained strict surveillance over the methods of distilling and rectifying.

Reversing the position, the question of quality must be allowed a prominent place in considering the very large proportion of insanity from intemperance in populous cities and large manufacturing centers. Here the poor quality of ardent spirits, the shocking sanitary condition of badly ventilated, overcrowded dwellings, the insufficiency of food, want of fuel and other like causes combine to render the drinker an exceedingly easy prey to the dreaded disease.

It is one of the inestimable merits of the Swedish system, that through the agencies of unselfish private associations, it has improved the condition of the laboring classes and thereby done away with many incentives to excesses.

Dr. Marvaud attributes inebriety to two principal causes; one is the "incredible activity and frantic struggle for gain," the other: "the insufficiency of food among the poor classes."† He is of opinion that the temperance question will be brought nearer to its solution by any efforts which would supply the poor with those aliments, whose want drives them to the use and abuse of stimulants. In our land of plenty, poverty and misery of the extreme type are found

* In another part of this sketch the working of the Swedish law is briefly described.

† "Cette activité incroyable, cette concurrence vitale effrénée." L'alcool, son action, son utilité et ses applications, &c. Paris, 1872.

only in very populous cities and large manufacturing centers; and in these localities the proportion of insanity from the cause in question is very high.

Not the least singular and significant feature of the letters contained in Appendix A, is the pronounced inclination on the part of a number of writers to counterbalance the seemingly inconsiderable effects of alcohol on the drinker directly, by making positive or suggestive statements as to the supposed effects of inebriety on the offsprings of besotted parents. As the question of heredity is entirely beyond the range of the present inquiry, only casual attention can be bestowed on such allusions. What we have to deal with here, are the *direct* effects of intemperance. We have seen, that in this respect, opinions of professionals and laymen move in extremes; and from this divergence of views in a matter capable of statistical verification, we may infer what must be the conflict of opinions on a subject, which has baffled the efforts of accomplished statisticians.

The result of Dr. Howe's inquiry (see page 5) seems to have been the only statistical guide to nearly all those who have written on the subject; for nowhere do we find more than learned generalizations, from which everything or nothing may be deduced, according to the inclination and predisposition of the reader. Excepting the English investigation, the only thorough and comprehensive inquiry into the effects of intemperance—so far as known to the compiler—is that which was instituted by the Danish government.* But the report of the Danish Statistical Bureau on inebriety contains but very meagre information on the point in question; and the dearth of such data argues inability on the part of the enquirers to obtain reliable statistics; for the report treats very extensively and minutely of the causes of pauperism and crime, and gives the proportion of drunkards to offsprings of drunkards throughout the kingdom from 1870 to 1880. Thus, for instance, the number of paupers from 1870 to 1880 is given, also the number of those whose dependence was caused by their own or their parents' intemperance, and furthermore the number of intemperate paupers, born of intemperate parents; the proportion of the latter to the former being 14 per cent. Had it been a task capable of accomplishment, the Statistical Bureau of Denmark, would no doubt have gathered similarly comprehensive statistics in reference to mental taints transmitted from

* Drikfaeldigheds Forholdene i Danmark, September, 1882.

drunken parents to their offsprings. As it is, there is but one asylum for idiots from which such data could be obtained. It is the private institution of G. Bakkehus and F. Mathisen. The proportion of idiots born of intemperate parents to the total number of admissions is as follows:

YEAR.	ADMISSIONS.	BORN OF INTEMPERATE PARENTS.
1871.	13	..
1872.	19	2
1873.	17	1
1874.	18	1
1875.	18	3
1876.	40	2
1877.	21	6
1878.	15	..
1879.	20	3
1880.	15	1
	196	19

The average proportion is about ten per cent. How does this compare with Dr. Howe's figure?

After this digression, we return to the showing of our table, to reiterate that seven per cent. is a fair average, that it agrees with the ratio in other countries, and is in harmony with the many internal evidences of decreasing inebriety in the United States.

If now, after having already made a number of unwarranted concessions to the opposing side, we should assume said proportion to fall short of the actual state of things by three per cent., we would certainly be justified in claiming that we have overdone fairness.

If, then, ten per cent. be accepted as a correct ratio, it would follow that the proportion of this class of insane to the population is one in over five thousand, since the proportion of insane to population is one in five hundred. That is to say, of the fifty-one million souls constituting the population of the United States in 1880, nine thousand two hundred have become insane by reason of intemperance; the total number of insane being 91,997.

What a difference between these figures and the estimates of temperance advocates!

As a matter of fact, the proportion of insanity caused by intemperance directly is seven per cent.; hence, of the entire population of our country, *six thousand four hundred and forty* have become insane by reason of excessive drinking.

ALCOHOLIC INEBRIETY, INSANITY AND BEER.

It would be an affront to the intelligence of the reader to demonstrate here, by reproducing the results of chemical analyses, the relative inebriating qualities of distilled liquors, wine and beer.

The most implacable enemy of King Gambrinus will readily admit that beer is the least intoxicating of these three kinds of beverages; but he will at the same time insist that beer-drinking leads to whiskey-drinking; and that beer, if used to excess, will produce insanity quite as surely, if not as speedily, as ardent spirits.

The slightest reflection must convince an impartial mind that if the use of beer had a tendency to create a craving for ardent spirits, the result would be the very reverse of what the revenue returns show to be the fact. The per capita consumption of distilled liquors would not then be on a decline, attended by an increase in the per capita consumption of beer, as has been the case during the decade ended in 1880. Indeed, a more powerful argument in favor of fermented beverages could not be conceived than is presented by the bare figures of the revenue returns, since they prove that beer, far from exciting an appetite for ardent spirits, has tended to largely diminish the per capita consumption of these very liquors; and whatever change has been effected in the drinking habits of the people, is due exclusively to the more general use of fermented drinks. Anyone familiar with life in the metropolis, knows that thousands upon thousands of former whiskey-drinkers now throng the pleasant halls and gardens, where music helps to stimulate that genial sociability and "Gemüthlichkeit" which beer invariably produces. With a taste for beer, the Americans have acquired also a knowledge of the art of recreation, in which they had theretofore been very deficient, and recreation is conceded to be a good preventive of insanity in many instances. In his "History of the Pennsylvania Hospital," Dr. J. Forsyth Meigs, reflecting the opinions of one of the most eminent physicians of our country, the late Dr. Kirkbride, says, in reference to recreations for the insane: "I will pause for a moment to ask whether these experiences of an intelligent medical observer of the value of amusements for the solace and cure of the insane, ought not to lead us to a higher appreciation of their value for the well.

Are not the Germans, as a nation, wiser than we, in the national habit they have formed of giving more of their time to entertainment and relaxation? They do no less work than we, of all kinds, mental and muscular, and yet appear to suffer less from insanity."

In this connection the compiler cannot refrain from quoting so acute and impartial an observer of events as the editor of *The New York Times*, who closed an editorial article on the celebration of the second centenary of the German immigration in the following words: "It would be difficult to compute the good that German immigration has done us in importing German music and German beer, and in the labor of the German immigrants as social missionaries, practically showing what was practically unknown in this country before they came, that it is possible on occasion to be idle and innocent."

Of similar utterances from equally good and trustworthy sources there is no dearth; they reflect the conviction, which is gaining more and more ground, that the German immigrants, who are under so many obligations to American genius and American institutions for their political and material well-being, have been the means of inculcating into the American mind an appreciation of Seneca's: "Now and then we should ease and refresh the mind with pleasures."

Prohibitionists will, however, attach but little weight to such general statements of fact; with the persistency that characterizes all their actions, they will, in spite of all that can be said, cling to their preconceived ideas, however seriously these may be in conflict with the sum of experience. For this reason it is deemed necessary to adduce statistical evidence of their errors.

The majority of Germans, not all of them, by any means, are habitual beer-drinkers. If the use of beer had a tendency to create an appetite for whiskey, the result would necessarily be a transformation of the drinking habits of the Germans, and the revenue returns would, we repeat, furnish the most reliable proof of such a tranformation. But there is, if not a more powerful, at least a more direct way of dispelling doubts on this question.

Appendix B contains a statistical report of five hundred cases of alcoholic inebriety, treated at the Inebriates-Home for Kings County, New York.*

* I am under great obligations to Dr. Blanchard, who has kindly caused such statistical information, as is not contained in Dr. Lewis T. Mason's valuable report, to be sent to me.

The report sets forth, among other facts, the nationality of the inebriates, the kind and quantity of drink ordinarily consumed, and the kind of drink to which the necessity for medical treatment is attributable. Tables III, IV, V and VI, being summaries of table II, show: That of the five hundred inebriates

338	were born	in the	United States,
92	"	"	Ireland,
27	"	"	England,
17	"	"	Canada,
13	"	"	Germany,
10	"	"	Scotland,
3	"	"	South America.

Kings county, according to the last census, had a population of 599,495 in the year 1880. The city of Brooklyn which is situated in Kings county, had a native population of 388,969 and a foreign population of 177,694; of the latter 104,291 came from Great Britain and Ireland, and 55,339 from Germany.

The proportion of native to whole population is	68.64
The proportion of immigrants from Great Britain and Ireland to whole population is	18.41
The proportion of immigrants from Germany to whole population is.	9.77
The proportion of native inebriates to the whole number of inebriates (500) is	67.60
The proportion of inebriates born in Great Britain and Ireland to whole number of inebriates is	25.80
The proportion of inebriates born in Germany to whole number of inebriates is	2.60

The tables also show that the necessity for medical treatment is attributable

To distilled liquors in	441	cases.
To distilled and fermented liquors	35	"
To fermented liquors	24	"
	500	

It is true, a large proportion of inebriates are reported as having been addicted to the use of both fermented and distilled liquors, but it is evident from the nativity of these persons that but for the newly created taste for beer, they would have been habitual whiskey-drinkers. With them beer plays the part of a mitigator of the habit. A man who drinks a pint of whiskey and ten glasses of beer daily, can scarcely be classed as an habitual beer-drinker.

Excepting two, the Germans who are included in table II. are, like so many of those coming from North-Germany, votaries of the old American drinking habit.

The tables further show that among the twenty-four habitual consumers of fermented drinks:

15	were born in	the	United States.
6	"	"	Ireland.
2	"	"	Germany.
1	"	"	England.

Of these twenty-four inebriates, twelve are reported to have suffered from complicating diseases or injuries; the majority of these had syphilitic diseases, of which it is well known that they frequently produce insanity, without the contributive influence of intoxicants. In truth, in the lists of assigned causes of insanity syphilitic diseases, sexual excesses and sensual vices occupy a prominent place.*

In order to enable the reader to judge as to what share fermented drinks have had in producing the affliction of these twelve inebriates, the compiler quotes from the fourteenth report of Dr. Lewis D. Mason, consulting physician to the Inebriates' Home in question: "The physician in all cases of Dipsomania, should look behind the mere symptoms of drink craving, and, as in Diabetes, in which the excessive thirst is merely symptomatic of disease, his remedies should be directed to the seat of the disorder. In some instances he will find Dipsomania to depend on a diseased condition that he can relieve. In other instances, as in some forms of head injuries, he can scarcely hope to cure the dipsomaniac; as in traumatic epilepsy, so in traumatic Dipsomania the prognosis must be extremely unfavorable."

The doctor cites instances in which the inordinate craving for drink was subdued by a removal of its exciting cause, the diseased condition of the patient. A case in point is that of a patient suffering from a combination of periodical dipsomania and stricture. Dr. Mason effected a cure of the latter disease, whereupon the excesses of periodical drinking ceased.

It must be kept in mind that persons are frequently confined in the Inebriates' Home, either of their wish and accord, or at the request of their relatives, simply to keep them out of the way of

* The annual report of the Asylum at Stockton, Cal. (1877-1878), under the heading of supposed causes of Insanity shows: Masturbation, 17. Intemperance, 24. This is an example taken at random.

apprehended harm. If this were not the case, it would be difficult to understand why No. 2, table IV., should be confined in such an institution. Here we have an editor, 22 years of age, who habitually consumed one quart of beer daily—one fourth of the quantity that the average Bavarian "Bürgersfrau" (house-wife) ordinarily consumes, without exhibiting the slightest symptoms of intoxication. In this case, mental over-exertion was doubtless the prime cause of the drinker's discomfiture.

The table of quantities consumed is very interesting.

In his work on Alcoholism, Dr. Baer dwells at length on the difference in the quantities of *distilled* liquors it takes to produce delirium tremens in different persons. Of ninety-six inebriates under his treatment, only thirteen had attacks of delirium tremens; the majority of the remainder had been accustomed to drink three fourths of a quart of whiskey or more daily, during periods of time, varying from six to twenty-five years; one patient ordinarily consumed from 1 to 1½ quarts daily for ten years, and another 2 quarts for eight years, without ever having an attack of delirium tremens.

In table IV. we find three inebriates who used either wine exclusively, or wine and beer. Two of the twenty-four inebriates had attacks of delirium tremens; one of these suffered from a concussion; the other, a female bartender, was addicted to the use of both wine and beer.

Recapitulating the showing of appendix B, we find that the native population, being 68.64 per cent. of the whole population, contributed 67.60 per cent. to the aggregate number of inebriates, that the Irish, English and Scotch, being 18.41 per cent. of the whole population, contributed 25 per cent.; and that the Germans, forming 9.77 per cent. of the entire population, contributed 2.60 per cent. The native population has one per cent. less than its proportion; Irish, English and Scotch have 6.59 per cent. more, and the Germans 7.17 less than their proportion. Of the thirteen German inebriates, only two were habitual drinkers of fermented liquors. Of the five hundred inebriates four hundred and forty-one owed their affliction to distilled liquors; thirty-five to distilled and fermented liquors, and twenty-four to fermented beverages exclusively. Of these twenty-four inebriates, twelve had grave complicating diseases or injuries; and only two, one who drank beer and wine, another who suffered from a concussion, had attacks of delirium tremens.

However unfavorably this showing may be construed, it surely cannot, from any rational point of view, be made to appear that beer has a tendency to supplant itself by creating an appetite for ardent spirits. The reverse is true, as must be admitted from the fact that no German habitual beer-drinkers, excepting the two last mentioned (one being a female bar-tender and the other a sufferer from injuries to the head), are among the inebriates; while twenty-two of the twenty-four inebriates addicted exclusively to the use of fermented beverages, belong to nationalities whose ordinary drink is *not* beer.

It is extremely difficult to determine what proportion of those inebriates who have had attacks of delirium tremens, ultimately become insane. Dr. Lewis D. Mason, * who, as has been said, is the consulting physician of the Inebriates' Home at Fort Hamilton, had the kindness to tell the compiler that it would be venturesome under any circumstances to give a statistical estimate. He did not hesitate, however, to say, that in cases where intemperance results in insanity, attacks of delirium tremens are generally—not always—the precursors. Now, of our 500 inebriates, 161 had attacks of delirium tremens, and of this number only two were habitual drinkers of fermented drinks. The proportion is 1¼ per cent. Of these two, we repeat, one was a female bar tender and the other suffered from a concussion.

How utterly untenable do all the exaggerated statements of prohibitionists appear in the light of such incontrovertible figures! In concluding this chapter, attention may again be called to Dr. v. Ziemsen's letter, showing that in Munich, where the per capita consumption of beer amounts to over 235 quarts, only 21 out of 10,000 patients suffered from alcoholism.

* Dr. Mason's opinion of malt liquors is expressed in his paper on "Alcoholic Insanity," in which, detailing the method of treating inebriates, he writes: "As a rule, I have found that when stimulants are indicated, the malt liquors are preferable to spirituous liquors. Bass' ale, Guiness' stout or lager beer, when a milder form is required. The value of malt liquors, in addition to their greater food properties, is due to their moderately stimulating qualities, combined with marked sedative or even hypnotic properties.

PAUPERISM.

The causes of this worst type of destitution are so multiform and complex that any attempt at generalization must lead to palpable fallacies.

Pauperism is found everywhere — in sterile countries of the North, where life has scarcely any charms for the laborer, where work is hard, and compensation small; as well as in fertile southern lands, where nature lavishes her richest gifts in greatest profusion upon many-headed indolence; where work is rarely more than a past-time and life an endless round of pleasure-seeking.

While pauperism presents the same pitiable aspect everywhere, it does not everywhere and at all times arise from the same set of causes; nor are the opinions regarding the latter more uniform. Economists are particularly unsuccessful in their efforts to agree upon some few general sources for what they term involuntary poverty, the result of economic and social conditions of different lands and times. If one would consult the works of economists of different nations on this subject, he would, passing from one language to another, have to unlearn to-morrow what he learned to-day. He would find that the dismemberment of large landed estates—remnants of the feudal system—has not accomplished for Germany what insular economists think it would for England; he would see that England's flourishing commerce, her mastery over the seas, and her industrial development, have no more prevented the growth of pauperism in Great Britain, than the French billions have in Germany; while in France, a country whose resources were drained almost to the bottom, he would find poverty decreasing.

One hears it asserted now and again, that the application of steam to manufactures and agricultural labor has created an army of involuntary idlers, paupers by force of circumstances, for whom the State ought to provide; that over-production and over-speculation, with their accompaniments of lock-outs, strikes, and financial crises, are at the root of this social evil. In short, there is no end of theories in regard to causes of involuntary poverty.

The question as to the sources of voluntary indigence is much more readily answered—if we allow prohibitionists to make the answer; which latter, in that case, would consist of the one single word: Intemperance.

Instead of investigating the correctness of this assumption, let us reverse the position, by enquiring whether or not poverty is a cause of intemperance. In doing so, we will follow a no less illustrious example than that set by the celebrated Liebig, who, in his "Chemische Briefe," says:

"In many places destitution and misery have been ascribed to the increasing use of spirits. This is an error. The use of spirits is not the cause, but an effect of poverty. It is an exception from the rule when a well-fed man becomes a spirit-drinker. On the other hand, when the laborer earns, by his work, less than is required to provide the amount of food which is indispensable in order to restore fully his working power, *an unyielding, inexorable law or necessity compels him to have recourse to spirits.*"

The same opinion is held by nearly all impartial medical writers, and forms, indeed, one of the guiding principles of the temperance agitation in Sweden and in the Netherlands, in England and in Germany.* Want of wholesome food, lack of pure air and water, and the total absence of home comforts and pleasures, render the life of the average laborer in many of the larger cities of these European countries almost unbearable. The social condition of these persons is recognized as the principal reason for their excesses in drinking, to which they frequently resort only to forget their misery, to subdue their craving for nourishing food, or to gain strength for their ceaseless toil.

Swedish humanitarians have, therefore, begun to build properly ventilated, comfortable dwellings for the laborers. They furnish food at prices covering cost of production only; they give with open hand all the means for public enjoyments, and institute reading rooms and popular theatres for the masses. The results of the Swedish efforts in this direction have only confirmed what was

* Roesch says: "The position of poor laborers, who are compelled to forego not only all the pleasures of life, but also the most needed nutriments, while working hard and incessantly, is a fruitful source of intemperance." Marvaud writes: "I say, then, to hygienists and legislators, that if they would successfully combat alcoholism, they must ameliorate the condition of the poor by giving them a sufficiency of wholesome food, and moderate doses of stimulants, mild and above all things unadulterated." Dr. Everts, Superintendent of the Cincinnati Sanitarium, in his "What shall we do for the Drunkard?" says: "It is probable, also, from the clinical history of drunkenness, that any cause of exhaustion of a special character, especially such as affect the brain and cord primarily; or a deprivation of nutritious and palatable food, on account of insufficiency, or bad cooking; or an inability to digest and assimilate food of a sufficiently stimulating character, becomes a predisposing cause of drunkenness."

theoretically known and practically tested, in a limited degree, many years ago in some of the populous cities of Europe. When the gin excitement was at its height in England, the question of combating intemperance among the working classes by measures tending to ameliorate the sanitary condition of workingmen's dwellings, was thoroughly ventilated, and many instances of the efficacy of such temperance efforts were brought to light. Dr. Southwood Smith cited a case in point: "In Lambeth Square, near Waterloo road, a population of 434 souls were huddled together. One person in five was diseased, and fifty and sixty per thousand annually died. The square was drained, water was made abundant, and used to carry away what formerly remained in cesspools. The change soon appeared. The mortality declined to thirteen per thousand. *The intemperate became sober*, and the disorderly well-conducted, after taking up their abode in these healthful dwellings."

Can a more powerful proof of the causality of poverty in connection with intemperance be conceived?

It has already been stated, according to the latest statistical reports from Sweden, that in spite of the stringent liquor laws, the consumption of ardent spirits has not decreased there until very recently; yet all the evils of intemperance, formerly complained of, had long ago been reduced to a minimum; and this favorable result is attributed, first, to the excellent quality of ardent spirits, and, secondly, to the ameliorated condition, materially and intellectually, of the workingmen.

Not only wholesome food for the body, but nourishment for the mind is sought to be procured for the workers, and in this latter respect, indeed, too much cannot be done in the interest of temperance. J. Leffort, in his oft-quoted work, "Intempérance et Misère," classifies idleness with the more prolific sources of intemperance—idleness in the sense of intellectual sloth. The action of idleness upon inebriety, he says, is so obvious that we do not venture to dwell at length on this point, fearing to bore the reader. Enough to say, that the man whose thoughts are fixed on nothing, whose mental faculties are without employment, whose limbs are idle, is by reason of his idleness more apt to yield to the caprices of his desires than he who is actively engaged in a task. The ennui which idleness engenders, readily leads to intemperance. Those who have studied the temperance question as it presents itself in England are agreed that the compulsory idleness to which the people of that country are condemned by the severity with which the so-called sanctity of the Sabbath is observed, contributes no little to excessive drinking. The relation which exists between the absence of all divertisements on that day, on which all work is suspended, and the frequency of excesses, has forced itself upon the attention of all tourists and publicists from the time of Bentham and Buret to our day. Fauchet, in his English sketches, arrives at the conclusion that the stricter the sanctity of the Sabbath is observed, the greater must inebriety necessarily become, because the intellectual sloth which seizes on the uncultured laborer when he has nothing to occupy or divert him, drives him to the bottle. As an illustration he cites Scotland, as being the most Puritanical country, but also the classic land of inebriety.

How forcibly do all these utterances remind one of the state of things in the tenement districts of our metropolis, and what enchanting visions of possible philanthropic achievements should they not reveal to our temperance advocates! The workers, knowing that a dreary day of compulsory idleness is before them, either supply themselves with a little store of intoxicants for that day, or compensate themselves beforehand by getting "jolly drunk." *

In Sweden, as well as in the Netherlands, this side of the question is well understood; hence we find temperance societies striving to emulate one another in their efforts to lift the laborer out of physical want and intellectual sloth. It is conceded there, as will be seen from a report in another part of this sketch, that neither laws nor exhortations will correct the evil, unless these measures be combined with efforts to substantially improve the condition of the laboring classes. Poverty, then, it seems, is recognized even by temperance advocates as a prolific source of intemperance. In reality intemperance is quite often the effect; poverty the cause, and pauperism the ultimate result.

The inquiry follows: Is such pauperism voluntary, or is it the result of the social and economic conditions of the land?

* The following item appeared in the New York *Sun* a few weeks ago: "Scotch inebriates continue to devote the hours between Saturday and Sunday, more than any other portion of the week, to drinking deep. Statistics just published show that between 6 A. M. on Saturday and the same hour on Sunday no fewer than 12,254 persons were arrested for drunkenness during 1882, while only 1,492 were seized by the police on Sunday and 17,977 during the rest of the week." It appears, then, the Scotch drinkers compensate themselves in advance for the weariness to which piety dooms on Sunday. It is a matter capable of statistical proof that in large German cities, where the places of enjoyment are doubled on Sunday, drunkeness and disorderly conduct is not more frequent than on week-days."

Professor Henry Fawcett explains his discrimination between voluntary and involuntary poverty by saying, in substance: Voluntary poverty is produced by the indolence, self-indulgence, or any other cause for which the individual who suffers is responsible; involuntary poverty includes all cases in which people become indigent through no fault of their own. Here he cites as examples children of extravagant parents, laborers thrown out of employment by a financial crisis produced by over-speculation of their employers, etc.

Now, who is to determine, in any case of pauperism, classed by officials as the result of intemperance, whether poverty, one of the prime causes of intemperance, as we have seen, was not originally brought on by just such a financial crisis?

Under date of February the 7th, 1884, the following cable dispatch was sent here from Leipsic, and published in the daily newspapers:*

"Widespread distress prevails among the working people of Saxony, owing to the dullness in manufactures and the paucity of employment. Seven of the sufferers committed suicide yesterday."

From the Saxon press, and from correspondences to papers outside of Saxony, it was learned subsequently that hundreds of unemployed laborers, driven to utter desperation by the gloomy aspect of the future, abandoned themselves to excesses in drinking, so as to forget their misery. Among these were said to be men who were formerly noted for their sobriety and correctness of conduct. Many of them will doubtless end in the poor-house; but will they then be voluntary paupers because of their intemperance? Is not poverty the cause of their excesses, just as it was the reason for the seven suicides, committed on one and the same day?

If indolence had brought on the intemperance of these laborers, it would be a different thing; but even then it would not be philosophical to argue that intemperance was the cause of pauperism—for indolence, the prime cause, is apt to lead to pauperism quite as surely without the aid of intemperance, as with it. Italy, the land of *il dolce far niente*, is the very paradise of lazzaroni; of beggars who are, in fact, neither more nor less than paupers, too fond of the golden flood of sunshine that flows from cloudless skies upon their

* See New York *Telegram* of same date.

happy land, to live in poor-houses; yet the Italians of all conditions, and particularly those of the lower classes, are exceedingly temperate and frugal in all their habits. In no European country is intemperance less frequent, and beggardom more prevalent, than in Italy. There the cause of pauperism is indolence.

If pauperism would ever, by honest and conscientious truth-seekers, be made the subject of comprehensive and fair inquiry in our country, it would doubtless be found that in voluntary as well as in involuntary indigence, intemperance is either the effect—not the cause—or merely one of the contributive causes. In contradiction to all the pet theories of prohibitionists, it would be ascertained that, aside from physical ailment, and the results of economic evils and scant natural resources, indolence and improvidence are the chief sources of pauperism, here as everywhere, now as of old.

"A vast number in every community are so constituted," says Fawcett, "that they would rather let others labor for them than labor themselves—they will not work unless compelled to do so." Now, if one of that vast number, who have nothing to lose, and care nothing for gain, spends for drink what he obtained by begging, and ends by becoming a drunkard—as some vagrants manage to do by simply emptying into their stomachs the remnants of beer found in kegs, piled up in front of saloons—he is not to be classed as a pauper by reason of his intemperance, but by reason of his indolence. If not a drop of liquor could be obtained in the land in which he lives, he would nevertheless be a pauper. In such a case, lack of self-respect and manhood are the causes of intemperance, as they are of indolence and vagrancy, and it matters little to which of these vices priority may be conceded. This should not be lost sight of in considering the relation which intemperance is said to bear to pauperism.

There is less indigence in the United States than in any other civilized country—less of both kinds of pauperism; and if the same confidence be reposed in the official statistics on this point, which prohibitionists place in the statistics of insanity, it must be affirmed that this social evil is rapidly decreasing. The census of 1870 puts the pauper population at 116,102; while in the census of 1880 only 88,665 paupers (inmates of almshouses and out-door poor) are accounted for. The argument of the advocates of prohibition, that pauperism, insanity and crime increase in proportion as inebriety does, might now be reversed with telling effect, if it were not unfair to ignore the explanatory article which precedes the statistics of pauperism in the census of 1880. It is stated there that, while the enumeration of the poor in institutions is very nearly correct, that of the out-door poor cannot absolutely be relied upon as including all those depending upon charity for the means of sustenance. But the same objection applies to previous censuses, and inasmuch as it may fairly be assumed that the method of enumeration has been improved since 1870, there is no reason why an actual decrease of pauperism might not be inferred from the difference between the figures of 1870 and those of 1880. The difficulty does not lie in accounting for more paupers than there actually are, but in not finding *all* of them; hence, when, with improved methods of search and enumeration, fewer paupers are found now than in former years, it stands to reason that in reality there are not as many now depending upon charity as there were in past years. For the purposes of this sketch it is immaterial, however, whether this view be regarded as correct or not. Certain it is, at all events, that the prohibitionists' theory in regard to increasing insanity, cannot be reconciled with the showing of the censuses with reference to pauperism.

What the compiler wishes to demonstrate by the collected data, is simply the relation which intemperance bears to pauperism. With this end in view, he selected the poor-house of Kings County, in which latter Brooklyn, the most populous city in the United States, next to New York and Philadelphia, is situated. No institution could afford a better test of the question mooted, so far as the position of prohibitionists is concerned, because this county has a large foreign population and an extensive manufacturing and shipping trade.

Through the kindness of the Warden of this institution, Mr. Murray—a gentleman thoroughly familiar with the system of public charity and the character of paupers—the history of six hundred and seventy-one indigent male persons, supported at the expense of the county, was obtained, and particular precautionary measures were followed to prevent erroneous classification. The Warden's observations bear out in every essential point the statistical statement contained in Appendix C, so that not the least hesitancy need be felt in accepting the figures as entirely correct.

Classified according to nativity, the six hundred and seventy-one paupers stand as follows:

United States	142
Ireland	332
Germany	124
England	35
Scotland	13
Sweden	5
Norway	4
France	3
Italy	2
Holland	2
Canada	2

and Switzerland, Denmark, Austria, Australia, Hungary, Finland and the West Indies, one each.

The cause of dependence was:

Physical disability in	457 cases.
Want of work in	99 "
Intemperance in	72 "
Vagrancy in	33 "
Senility in	10 "

An examination of Appendix C will convince any impartial person that under the heading of physical disability no paupers are included whose disease or injury would not incapacitate any man to earn his own bread. Those classed as vagrants are incorrigible drones, who either voluntarily or compulsorily take up quarters in the poorhouse during the winter, and resume their aimless wanderings as soon as fair springtime returns. Those classed under the heading of senility (see Table VII.) are men of between eighty and ninety years of age. Many of the paupers from foreign lands bring their physical disabilities with them, thus forming an infinitesimal counterweight—light as a feather—to the powerful aid we receive from Europe in money, muscle, skilled labor and brain. This fact, of which our government had to take cognizance quite frequently, and which until very recently formed the subject of diplomatic negotiations with a number of foreign governments, accounts for the disproportionately large number of paupers born outside of the United States.

Of the seventy-two paupers whose dependence is reported to have been caused by intemperance,

38 were born in	Ireland.
28 " "	United States.
3 " "	Germany.
2 " "	England.
1 " "	Scotland.

The proportion of pauperism caused by intemperance is 10 74 per cent. Among the female paupers there are very few intemperates; but in the female department and nursery of every poorhouse there are, of course, a number of inmates whose indigence was caused by the intemperance of husband or father, and in estimating the number of persons made dependent upon public charity by drink, this fact must be taken into consideration. As it is extremely difficult, not to say impossible, to obtain correct data on this point from the institution here referred to, the compiler adopted the proportion which appears from the Danish statistics of pauperism,* adding 5 per cent. for females and 8.03 for children, so that the total proportion of pauperism caused by intemperance is raised from 10.74 per cent. to 23.74 per cent. Hence we have four intemperate paupers to every ten thousand of the entire population.

Here, again, the most unfavorable showing has been taken as a basis of calculation; for it would surely have been more appropriate and advantageous to adopt the figures of the census of Massachusetts for 1875, compiled by C. D. Wright. The table of causes of pauperism in that census contains these figures:

Intemperance	473
" of husband	18
" " father	40
" " mother	1
" " parents	52
	584

The total number of paupers was 4342; hence 13 per cent., instead of 23 would have been the proportion, if this basis had

---

* Year.	Number of Paupers.				Number of Paupers whose Dependence was Caused by Intemperance, Directly or Indirectly.				Proportion of Pauperism Caused by Intemperance.			
	Men.	Women.	Children.	Total.	Men.	Women.	Children.	Total.	Men.	Women.	Children.	Total.
1871	2508	2004	1700	6212	921	373	508	1802	36.72	18.61	29.88	29.01
1872	2273	1929	1699	5901	806	346	446	1598	35.46	17.94	26.25	27.08
1873	2105	1822	1668	5595	760	341	488	1589	36.10	18.72	29.26	28.40
1874	2123	1819	1591	5536	848	345	461	1654	39.94	18.97	28.92	29.83
1875	2210	1851	1636	5697	907	346	464	1717	41.04	18.69	28.36	30.14
1876	2253	1871	1633	5757	933	361	491	1785	41.41	19.29	30.07	31.01
1877	2737	2027	1792	6556	1125	400	569	2094	41.10	19.73	31.75	31.94
1878	3228	2261	1911	7400	1251	409	598	2258	38.75	18.09	31.29	30.61
1879	3385	2227	1950	7562	1349	426	584	2359	39.85	19.13	29.95	31.20
1880	3502	2358	2172	8032	1367	447	663	2477	39.03	18.96	30.52	30.84
Total	26324	20169	17755	64248	10267	3794	5272	19333	39.00	18.81	29 69	30.09

From Drikfældigheds Forholdene i Danmark (1882) p. 64.

been adopted. But it may be better to overdo fairness than to expose oneself to the opposite reproach.

Of the seventy-two persons rendered indigent by intemperance,

52.78	per cent. were born in............	Ireland.
38.89	" " "	United States.
4.16	" " "	Germany.

The drinking habits of intemperate paupers are generally of the worst kind; these weak-willed people will drink anything and everything, with a decided preference for the strongest kinds of beverages; but not one of them can be classed with habitual beer-drinkers. This averment will be better understood in the light of a conversation the compiler had with one of the German paupers, a man past three score years of age, upon whose emaciated countenance misery had written her rugged signature in countless wrinkles and furrows. Being addressed in his native tongue, and in a manner betraying at least interest, if not sympathy, the old man became rather more communicative than such people usually are, and in an unmistakably contrite mood and a spirit of self-accusation, related the story of his life. His father's hovel, he said, stood on a bleak and barren heath, far up in the north of Fatherland. There he grew up to manhood, following the occupation of all his neighbors—that of cutting turf from extensive peat-bogs. Coarse black bread, bacon, and spirits, distilled from cereals or potatoes, formed his regular diet. After his father's death he married—according to some economists the greatest crime a poor man can possibly commit—and then continued in company the same dreary drudgery of life he had before led singly. Twenty years later his wife and only son died in rapid succession. In his grief he abandoned himself to excessive drinking, and soon became a shiftless idler. The "emigration fever" raged in his neighborhood at that time, and one fine day, after having disposed of the remnant of his household goods, our man set out on a journey to Hamburg, where he embarked for America. Here he obtained employment, and led an orderly life, becoming accustomed to drinking lager beer, a beverage he had not known in his old home. But "hard times" set in; he lost his employment, used up the money he had hoarded for just such an emergency, and soon found himself penniless. Easily moved to despondency, he became moody, and relapsed into the old habit of drinking to intoxication, selling one article of clothing after the other to procure ardent spirits. He would not touch beer in such moods. To use his own words: "Beer and wine are drinks for happy men, who wish to chat and laugh and be merry. I bought the worst kind of whiskey, because I knew it would make me drunk quickest, for what I wanted was, to get away from myself." If he could not have obtained whiskey, he would have committed suicide, he thought. Thus he continued drinking until he had become a confirmed drunkard and a pauper.

That he should have felt himself placed before the alternative of intoxication or self-destruction is not unnatural. Do we not read every day of men who end their lives, when starvation stares them in the face? Temperance advocates would doubtless say that it were better to die, even of one's own hand, than to become a drunkard; but humanity readily suggests an excuse for the wretch who becomes a drunkard in preference to becoming a self-murderer.

Our German pauper's case, so far as the kind of drink is concerned, may be taken to be typical of all those cases in which intemperance is assigned as the cause of dependence. Habitual beer-drinkers are not, consequently, represented in the body of intemperate paupers. The simple figures, showing the percentage of Americans, Irishmen and Germans among the intemperate paupers, are intensely eloquent on this point.

We are told that if it were not for intoxicating drinks, the people would be freed from three-fourths of the taxes they now pay. Let us examine this assertion statistically so far as insane and paupers are concerned.

The insane population of the United States numbered, in 1880	91,997	
Seven per cent. of that number (made insane by drink) is	6,440	
The average weekly cost of maintenance of each insane is	$1.50	
The pauper population of the United States in 1880 numbered	88,665	
Twenty-four per cent. of this number (being intemperate paupers) is	21,279	
The average weekly cost of maintenance of each pauper (according to the last report of the Commissioners of Charity of Kings Co.) is	$1.20	
The annual cost of maintaining the whole number of persons made insane by drink was 6,440 × 52 × $1.50		$502,320
The annual cost of maintaining the whole number of persons rendered indigent by drink was 21,279 × 52 × $1.20		1,327,809
Total		$1,830,129

In 1880 the United States Treasury Department received taxes from the manufacturers of distilled liquors to the amount of....................................	$55,919,119.18
From manufacturers of fermented beverages.......... ..	12,346,077.26
Total....................................	$68,265,196.44

The excess of revenues from liquor tax (leaving aside retailers' taxes, amounting to over $5,000,000) is $66,435,067.44. The fact that this sum is paid into the United States Treasury, while the insane and poor expenses are covered by State, county or municipal taxation, does not change the aspect of things, for the money required to liquidate the national debt and defray the cost of maintaining the National Government would, if there were no liquor taxes, have to come out of the taxpayers' purses, either directly or indirectly.

Besides, this amount does not constitute all the revenues derived from these sources. In every county, city and town, where the laws do not prohibit the sale of liquors, the privilege to sell such drinks is heavily taxed. Although it is impossible at present to show how much money flows into county and municipal treasuries through such taxation, an approximately correct idea may be formed from the fact that the revenues of nearly all the large cities, derived from excise duties, amount to sums, equal to all the expenses incurred in providing for the *entire* insane and pauper population in institutions there. The excise duties of the city of Brooklyn alone, for instance, amounted for one year to $230,250. The expenses for the maintenance of the entire insane and pauper population of the whole county, in the last year, amounted to $236,285.04.

The majority of penal institutions are either wholly or partly self-supporting, and hence the taxpayers have to contribute comparatively very little to their maintenance. But let us assume no penal institution at all to be self-supporting, and then let us see what crimes and offenses of all kinds would cost the taxpayers—not crimes caused by intemperance alone, but crimes and minor law-violations generally.

The census of 1880 fixes the number of prisoners in all the penitentiaries, county jails, city prisons, workhouses, military prisons, and insane hospitals, and of those otherwise detained, at 59,255. The weekly cost of maintaining one prisoner would, on account of the greater number of paid employees needed in penal institutions, be greater than that of maintaining a pauper. If we assume the weekly cost of each prisoner to be $1.50, we get $4,621,890 as the total annual cost of maintaining all the prisoners in the land. Let us go still further:

The cost of supporting the entire insane population of the United States, 91,997 persons, at an average cost of $1.50 per week, is.	$7,175,766
The cost of supporting the entire pauper population of the United States, 88,665, at an average of $1.20 per week, is	5,532,696
The cost of supporting the entire criminal population of the United States, 59,255 persons, at an average of $1.50 per week, would be, if penitentiary were not self-supporting.	4,621,890
Total..........	$17,330,352

Thus it is evident that, while alcohol is the cause of only seven per cent. of insanity and twenty-four per cent. of pauperism, it pays in one year to the national treasury alone, not to mention the enormous sums of local taxes, nearly fifty-one million dollars more than it costs to support the *entire* insane, pauper and criminal population of the country. Far from freeing the taxpayers from three-fourths of the taxes they now pay, prohibition would, even if it were as practicable, as moral and as logical as it is the reverse, impose upon the taxpayers heavy loads of new duties.

The moral side of the question will be considered in one of the succeeding chapters.

CRIME.

"You have two hundred and sixty criminals, commonly styled long term convicts, under your charge; how many of them, do you suppose, have committed the crime for which they are now being punished, under the influence of intoxicating liquors."

This question was asked of Mr. Green, Warden of the Penitentiary of Kings County, a gentlemen whose chief intellectual characteristic is soundness of judgment, wedded to great power of observation. His answer to the compiler was: "It is not a matter of mere supposition when I say, that not even five per cent. of the criminals can be said to have committed their offense in consequence and by reason of intoxication. I do not mean to say that criminals are as a class more temperate then law-abiding people; but I do assert that intoxication is very rarely the cause of those crimes which affect the security of property; while capital crimes are much more frequently the out-come of cold-blooded premeditation—whether a murderous disposition, greed of gain or turbulent passions, such as jealousy, be the primary motive—than of intoxication. In my opinion women are the bottom of a great number of crimes. Nearly every forger, burglar and highwayman in this institution is regularly visited by a woman—usually a gaudily dressed creature who displays an uncommonly warm affection for the object of her visits. It is to provide such creatures with finery, that burglaries, forgeries, robberies and like crimes are most frequently committed. Drink has little to do with it; in fact, sobriety, steadiness of nerve and no little mechanical skill are indispensable to the perpetration of a successful burglary, and there are numerous crimes that exact extreme clear-headedness in the perpetrator. Any one who through his official position becomes familiar with the lives of criminals, will tell you that drunkenness is very rare among the most dangerous classes of offenders. Of the four hundred and five short-term convicts quite a number are addicted to the excessive use of intoxicating liquors; but their intemperate habit is but one of a number of vices, all of them the result of a lack of moral and intellectual training. With whiskey or without it, such persons' perverted sense of right and wrong would lead on to crime under any circumstances."

At the request of the compiler, Mr. Green made an attempt to sift the histories of all the cases under his charge, so as to arrive, if possible, at an exact statistical estimate of the proportion of crimes committed through and by reason of intemperance. Failure attended the trial. For inasmuch as the most fervent hopes of all convicts are fixed on the possibility of securing the exercise of executive clemency in their behalf, they at once suspected that the inquiry into their antecedents was instituted with a view to abbreviating their term of imprisonment, and knowing that a pernicious sentimentality prevails in reference to victims of drink, they were but too prone to plead inebriety, or even alcoholic insanity in extenuation of their crime. Efforts made by other truth-seekers in other parts of our country have yielded few satisfactory results, viewed from a statistical standpoint. Two among a few exceptions are the investigations instituted by the State authorities in Massachusetts, and that conducted by Rev. John Ruth, Chaplain of the Penitentiary for the Eastern District of Pennsylvania, the result of which latter, as will be remembered, was a complete refutation of the theory that intemperance is accountable for three-fourths or one-half of all crimes. The principal cause of crime was by this investigation clearly demonstrated to be the lack of a trade education, and this was accepted as correct by publicists, who can not in any conceivable manner be suspected of a leaning towards the liquor interests. As one of many instances, we quote the words of Charles F. Thwing, first published in the Christian Union of the 30th of October, 1878, and reproduced in the Annual Report of the New Jersey Statistical Bureau of Labor and Industry.

"The statement is constantly made that intemperance is the cause of nine-tenths of the crime committed in this country. But an examination of the reports of the prisons of the United States proves that the influence of rum in exciting to crime is greatly exaggerated. That its influence in promoting lawlessness is great —very great—cannot be doubted, but that it is as great as usually represented, cannot be proved. To the lack of a trade education must be contributed much of the crime which is commonly attributed to liquor drinking."

There is no doubt much truth in this. A powerful incentive to crime, much more potent than intemperance or any other cause, is destitution, and the lack of a trade education frequently conduces to poverty; but the reverse does not insure immunity from that evil which Diderot regarded as worse than crime. Individual propensities and the social and economic conditions of communities may completely counteract the advantages of a trade. One is but too apt to fall into error if he places too much stress upon externals, treating the moral constitution of the criminal, the qualities born with him, as of minor importance. Cain became a murderer without any of the external incentives which students of social science of our day classify as causes of crime.

To arrive at a correct judgment, it is necessary to take into consideration, everything, external and internal—moral defects in the individual, his surroundings, his education, his physical and intellectual capacity on the one hand, and the condition of the community in which he lives, the method of administering justice and the penal system on the other. These requisites are such that a large corps of investigators would be needed to give us more than generalizations.

As far as intemperance is concerned in the causing of crime, there seems to be a firm conviction on the part of many prison officials, that drink is at the root of such offenses only as belong almost exclusively to the jurisdiction of minor judicial courts.

This view is sustained by the observations of many competent investigators. In his highly interesting and valuable work on Alcoholic Inebriety,* Dr. Joseph Parrish, of Burlington, N. J., writes on this subject:

"Dr. Arnold, of Baltimore, speaks of inebriates thus:

'Inebriates do not form that class of people who plan and carry out schemes of villainy and corruption, in high and low places; nor are they usually found on the list of professional criminals who figure in our courts of law. Besides, it is notorious how often criminals try to mitigate the heinousness of their offences by attributing them to the effects of alcohol.'

It should be borne in mind, also, that the very habit of intoxication disqualifies persons from committing some crimes. The habitual and excessive use of intoxicants promotes timidity, incautiousness and inefficiency, and failure is the almost invariable result of attempts to commit certain kinds of crime by those who indulge in intoxicants. An expert was some time since employed to search the records of crime in a neighboring State, with a view of ascertaining from official sources the number of persons convicted of murder during the past hundred years, with the causes, penalties, etc., etc. After a careful and painstaking examination of court and prison records, it was reported *that less than three per centum of such crimes could be traced to the use of intoxicating liquors.* Upon this disclosure being made, it was repeated to a certain temperance advocate of the same State, who confirmed its accuracy by saying that he had caused a similar investigation to be made, with the same result, but added that he hesitated to make it public, because it would deprive advocates of temperance of a cogent argument in behalf of the cause. Pursuing the same line of inquiry from time to time, it fell in my way to ask a very worthy chaplain of a penitentiary how many of the several hundreds of convicts under his care could connect their crimes with the use of intoxicating drinks. His reply was, that from direct personal knowledge of the history of each prisoner, he believed they were all guilty of vices—such as gambling, profanity, falsifying, tobacco chewing, smoking to excess and lewdness, etc.—but to which of these vices their particular crime was to be attributed he could not tell, but that it would be about as easy and fair to trace it to one as to another; and he added: 'Those whose crimes are the direct result of intemperance are very few. *I do not know of one.*' It would be more philosophical to go behind and beyond them, to the source from which they all spring—namely, a depraved physical and moral nature. Being children, all of the same stock, their conduct and behavior originated in one common source, and it takes either line that is indicated, in accordance with the direction of certain physical tendencies.

The Hon. Richard Vaux, of Philadelphia, distinguished as a penologist of rare powers and opportunities for observation, writes me as follows·

'I do not consider intemperance, as it is called—inebriety, the use of intoxicants—as a crime-cause. If this were so, all inebriates would be criminals. Now, the fact is, that criminals are made so by other causes; and they, like the rest of mankind, use intoxicants or *do not* use them. It is now forty years since I have been an Inspector of the Eastern State Penitentiary in this city, and I have no

* "Alcoholic Inebriety from a Medical Standpoint." By Dr. Joseph Parrish. Philadelphia. P. Blakiston, Son & Co. 1883.

hesitation in saying that intemperance—the use of intoxicants habitually, or to excess—is not a crime cause. I think it can be said, that about one-half of those convicted of crime are total abstainers. Of the four hundred and thirty-three (433) prisoners received into our Penitentiary in 1881, but twenty-six (26) were intemperate. Mr. Cassiday, our Warden, who has been in the service of this prison for twenty years, gives his experience in support of these views of crime-cause. I know it is a sort of fashion to talk about our prisons filled with the victims of intemperance, but the figures do not support this general and sweeping assertion.'

In confirmation of the same views, I am furnished with the following from the accomplished General G. Mott, late keeper of the New Jersey State Prison, at Trenton:

'I am decidedly of the opinion that our Penitentiaries are not filled with those who trace their crimes to intemperance; that class fill our common jails, lockups, and Houses of Correction. A person sent to a Penitentiary, no matter for how short a time, for a violation of the law, perhaps committed in the heat of passion and while under the influence of liquor, is branded a criminal; thinks society has injured him, and when he gets out may join the criminal class, as he says, "to get square;" but he must keep sober if he expects to get in with the expert. The majority of criminals who fill our Penitentiary are primarily of a criminal mind, born so, and brought up to prey upon the general community; but they are not habitual drunkards, nor do they associate with that class. Not so themselves, because, to be an expert, they must keep their heads cool and their wits about them; and their associates must do the same, as they know there is no dependence on a drunken man; for, when in that condition, he may let something drop that, perhaps, will lead to the failure of their plans and the probable detection of the principals.'

The following is also contributed from the Maryland State Penitentiary:

'Out of five hundred and thirty-four (534) convicts in November, 1881, there were strictly temperate one hundred and seventeen (117); moderate drinkers, two hundred and forty-two (242); occasionally intemperate, one hundred and seventy-one (171), and habitually intemperate, four (4).'

From Mr. John C. Salter, the successful Warden of the State Penitentiary at Chester, Illinois, I learn the following:

'The popular sentiment seems to be that a criminal must necessarily have been a drunkard. That this class are frequenters of saloons, and are more or less slaves to appetite for strong drink, as they are to other vices, cannot be denied. The large proportion of criminals, such as burglars, forgers, counterfeiters, need clear brains, steady nerves and quick perceptions to successfully carry out their plans, which would be impossible under the influence of intoxicating drink. I am more and more convinced that the causes of crime go away back in the history of the criminal, even outside of his own life, coming down from generation to generation, visiting the iniquities of the fathers upon the children. Lack of home influence, throwing boys and girls of tender age out upon the charity of the world; lack of the discipline of education, the haste to get rich and the false standard of greatness are causes that have done much toward filling our jails and penitentiaries with those who, under more favoring winds, would have found shelter in a friendly harbor.'

The apparent discrepancy between the commonly accepted belief that at least two-thirds of all the crimes are due to intemperance and the actual facts, as derived from institution statistics, may perhaps be accounted for thus: The offences for which persons are sent to houses of correction, county jails and lock-ups are largely attributed to strong drink as an *exciting* cause, while the more grave offences are punished by commitment to penitentiaries. Also, the commitments to the common places of detention are counted over and over again, and the evil is made to appear, as we shall presently see, much more formidable than the facts really justify. Vagrancy is an offence that does not find its way to penitentiaries, and yet it occupies a conspicuous place in the common jail records. Vagrancy is often associated with drunkenness, but not always as cause and effect. Pauperism and vagrancy are usually associated with a low and depraved physical and moral constitution. In many cases the tendency is to despondency, and despondeucy is frequently an exciting cause of intemperance. If, therefore, vagrancy is counted as crime, and every vagrant who drinks is counted as intemperate, it can be readily seen how so large a percentage is given to intemperance as a crime-cause. So, if intoxication is counted a crime, and, to use the police nomenclature, if "drunk and disorderly" is a title attached to every commitment for intemperance and vagrancy, the showing in that direction must necessarily be exaggerated. And yet it is just about in this careless manner that the police records are frequently kept. A scientific nomenclature is unknown to the law, while the docket of a police justice cannot be more than a transcript of the justice's own ideal of what is, and what is not crime, or disease. Crime usually has its source in the mental or moral constitution. The desire for alcoholic beverages is generally a physical desire, an animal lust, and has but a distant, if any, relation to what is recognized as the moral character."

The experiences of prison officials in nearly all civilized countries confirm these views; at all events there is no statistical proof to the contrary, and unless that can be adduced, on either side, all discussion on the subject will necessarily have to be based on just such opinions as have just been quoted, or their opposite. There is, however, one feature of this question which is capable of statistical verification, and which offers many inducements to thorough research, namely, the relation which fermented and distilled liquors bear to those law-violations that are held principally to be caused by intoxication. Difficult as the task of establishing these relations appears to be at first sight, it is in reality quite easily accomplished if the proper authorities take it in hand.

Acting upon instructions from Mayor Low, the Excise Commissioners of Brooklyn caused to be prepared a statistical report of the number of every class of saloons in each precinct, the number of arrests and the population. And this report shows most conclusively what has been demonstrated in this sketch by scores of authentic examples, *i.e.*, that the use of beer practically effects the objects of a rational temperance movement. The statistical table, being so full of interest and importance, surely deserves to be reproduced here in full.

Precinct.	Estimated Population for 1883, by Precincts.	Arrests for Intoxication.	No. of Saloons for 1883, by classes. First.	Second.	Third.	Total Number of Saloons.	Percentage of Arrests for Intoxication to Population, 1883.	Pro rata of Arrests for Intoxication to total number of Saloons, 1883.	Population to each Liquor Saloon (first and second classes).	Population to each Beer Saloon (third class).	Population to each Saloon of all kinds.
First...... ..	43,900	1,471	151	18	14	183	$3\frac{3}{10}$	8	260	3,130	240
Second........	31,100	1,138	148	8	8	164	$3\frac{8}{10}$	7	200	3,900	190
Third.........	49,000	1,602	119	19	20	158	$3\frac{8}{10}$	10	355	2,450	317
Fourth........	63,500	1,001	125	31	12	168	$1\frac{6}{10}$	6	407	5,300	380
Fifth.........	54,500	860	180	21	97	298	$1\frac{8}{10}$	3	270	557	180
Sixth.........	72,500	393	155	10	216	381	$\frac{5}{10}$	1	440	335	199
Seventh......	33,000	430	105	5	45	155	$1\frac{8}{10}$	$2\frac{8}{10}$	300	755	213
Eighth........	25,200	368	105	3	22	130	$1\frac{5}{10}$	$2\frac{8}{10}$	233	1,150	194
Ninth.........	44,500	254	53	18	8	79	$\frac{6}{10}$	$3\frac{2}{10}$	627	2,470	563
Tenth	62,800	742	117	17	19	153	$1\frac{2}{10}$	$4\frac{8}{10}$	470	3,305	410
Eleventh......	32,400	1,627	163	3	29	195	5	$8\frac{3}{10}$	195	1,117	166
Twelfth	25,200	293	63	10	15	88	$1\frac{2}{10}$	$3\frac{3}{10}$	345	1,680	287
Thirteenth....	55,200	445	119	10	154	283	$\frac{8}{10}$	$1\frac{8}{10}$	429	358	192
Third Sub.....	15,100	1,705	55	1	5	61	$11\frac{8}{10}$	28	270	3,020	247
Fifth Sub. ...	16,500	182	28	5	6	39	$1\frac{7}{10}$	$4\frac{6}{10}$	500	2,750	460
Eighth Sub...	5,800	198	22	1	13	36	$3\frac{4}{10}$	$5\frac{5}{10}$	252	446	161
Ninth Sub.....	15,600	132	60	4	18	82	$\frac{8}{10}$	$1\frac{8}{10}$	244	866	190
Total.....	645,800	12,841	1,768	184	701	2,653	Av. 2.	Av. 4 8-10	Av. 331	Av. 921	Av. 244

In saloons of the first class ardent spirits and fermented liquors are allowed to be sold; in saloons of the third class only beer is sold. Licenses of the second class are granted to storekeepers who sell ardent spirits by the measure, in quantities not exceeding five gallons.

The greatest number of beer saloons are in the Sixth and Thirteenth Precincts, embracing the greater part of that portion of Brooklyn which is sometimes vulgarly styled "Dutchtown." In these two precincts the number of arrests for intoxication was smallest. From the Sixth Precinct, with a population of 72,500 souls, and 216 beer saloons, only 393 arrests for intoxication are reported; in the Thirteenth Precinct, with a population of 55,200 souls, and 154 beer saloons, 445 arrests were made for the same cause; the proportion of arrests to population being $\frac{1}{2}$ per cent. in the former, $\frac{4}{5}$ per cent. in the latter precinct. In the Third Sub-Precinct, with a population of 15,100, 55 first-class saloons, and only 5 beer saloons, 1,705 arrests were made for intoxication; the proportion of arrests being $11\frac{3}{10}$ per cent. The First Precinct, with 183 saloons, 14 of which are exclusively beer places, shows 1,471 arrests, or $3\frac{3}{10}$ for every hundred of the population. Nearly the same numerical relation between first-class and third-class saloons exists in the Second and Third Precincts, and the proportion of arrests is also nearly the same in the three precincts. The Ninth Precinct, with 79 saloons, of which 8 only are beer places, and a proportion of arrests of $\frac{6}{10}$, the population being 44,500, presents a very significant showing, compared with the Third Sub-Precinct and First, Second and Third Precincts. Beer does not, however, enter into the question there; but the drinking habits and condition of the people do, so does the quality of drink consumed; three factors which, as has repeatedly been demonstrated, are just as decisive in the matter of inebriety as quantity is. The Ninth Precinct with its seventy-nine saloons and 44,500 inhabitants shows 254 arrests, while in the Third Sub. with only 61 saloons and 15,000 inhabitants 1,705 arrests were made. The former precinct is principally inhabited by well-to-do people who can afford to purchase good stimulants, while the latter is crowded by poor persons whose surroundings, mode of life and habits inevitably tend to develop the germ of excesses. There, want of wholesome food, of proper shelter, of all the comforts of life combine to drive men and women to excessive drinking; but worst of all, the ardent spirits they can afford to buy are, in consequence of their cheapness, of a most execrable quality.

A more striking illustration of the vast difference between the effects of good and pure, and those of adulterated and impure ardent spirits cannot be conceived.

The Swedish laws, regulating the manufacture of distilled spirits, secure the drinker against the effects of adulterated and insufficiently purified ardent spirits; and the Gothenburg system encourages the use of beer. The result is, that, in spite of only slightly diminished consumption and increased population, the number of arrests for intoxication has decreased steadily in all the larger cities. In the official report of the German committee before-mentioned we find the following table relating to the city after which the Gothenburg system is named.

Year.	Population.	Delirium tremens.	Arrests for Intoxication.
1877	153,528	436	4,548
1878	161,722	264	3,740
1879	163,040	227	3,648
1880	167,868	241	3,744
1881	174,706	234	3,537

These figures are worth all the temperance tracts ever published.

If the Excise Commissioners of the City of New York would cause a report to be prepared on this subject, it would be found that the proportion of arrests in what might be styled beer districts is quite as small there as it is in Brooklyn. The proportion of arrests to population in the 6th and 13th Precincts—$\frac{5}{10}$ and $\frac{8}{10}$ respectively—would, it is claimed, be still lower, if these localities were inhabited by habitual beer-drinkers exclusively; for it is said to be a matter of police record that among those who are arrested for intoxication in said precincts, not ten out of a hundred are Germans. This calls to mind the fact that on the days of great German festivals, when thousands upon thousands of people congregate in gardens and halls for pleasure's sake, and when beer is consumed in enormous quantities, arrests for intoxication or disorderly conduct on the part of the participants of such festivals are scarcely ever heard of.

The festival of the Suabians, held in 1883 at a park on the outskirts of Brooklyn was participated in by twenty-two thousand persons; 490 kegs of beer and about 250 gallons of wine were consumed—and not one arrest was made on the ground.

The festival of North-Germans held in 1883, at Union Hill, attracted a concourse of men, women and children, numbering 48,000 heads. In all, five arrests were made for disorderly conduct.

Dr. Bowditch says in reference to this subject: The extraordinary difference between the percentages of crime perpetrated by Germans and Irish is a peculiar fact, to be interpreted partly by the differences in the temperaments of the people, but still more I am inclined to believe by the difference in the liquors used by the two. I cannot but think that if the Germans were to drink rum and whiskey as the Irish do, a much larger proportion of crime would be found among them than now, for whiskey does not, so far as I know, affect a German body differently from an Irishman's body. I would likewise suggest the following proposition. Take away the whiskey from the Irishman and persuade him to use lager beer or Bavarian ale, and perhaps you will take from him a good deal of his pugnacity, and he will be less frequently drunk. * * * In truth I do not remember to have ever seen a German reeling home intoxicated, or sound asleep on some doorstep, evidently narcotized as the Yankee or Irishman is likely to be by some violent liquor." One of Dr. Bowditch's correspondents, Mr. Y. G. Hurd, Superintendent of the House of Corrections at Ipswich, Mass., writes: "I visited the beer gardens on Sunday, (in Chicago) to see how the Germans spend the day. There was a band of music, a dance floor, with seats and tables like our New England pic-nics in a beautiful grove, and lager in such quantities as I had never conceived. Everybody, old and young, drank and seemed to continue to drink during the afternoon, but lager was the only beverage—no liquors, no drunkenness, no fight, no disorderly conduct. The young men and maidens were merry and danced; the elders drank and talked with the gravity and dignity becoming to the respectable German."

The foregoing statistics and opinions fully establish the relation which beer bears to that class of law-violations which are usually attributed to intemperance.

USE AND ABUSE.

We have seen that the effects of the abuse of intoxicants are unwarrantably exaggerated; that, considered from an economic standpoint, and in their relation to the welfare of society, they are counterbalanced more than a hundred-fold by the fiscal advantages which the community derives from the use of intoxicants. If, therefore, pecuniary considerations only were to decide, the conclusion would be inevitable that doing away with the use of inebriating liquors—if such a thing were possible—might perhaps save an infinitesimal fraction of the population from the insane asylum, the poor-house or the jail, but that it would be a great material loss to the community. True, the revenues derived from this source are, at present, to a great extent exceptional, having grown out of the necessities of war; but even so, the duties which the liquor traffic is made to pay under a rational license system exceed by far the amount of loss sustained by the community in consequence of the abuse of intoxicants. And this is but a narrow view of the matter.

The manufacture of distilled and fermented spirits is closely connected by many important ties with agriculture, commerce and industry. The capital invested in distilleries, breweries and malt-houses alone amounted, in 1880, to $137,428,171. The production of barley and hops, important branches of our agriculture, depends almost entirely upon these industries, not to speak of the hundreds of branches of mechanical industry (such as manufacturies of refrigerators, of brewing and distilling vessels and implements, etc.,) which are dependent directly or indirectly upon the liquor traffic. The destruction of this traffic, it will readily be understood, would entail losses which could not but affect the agricultural, industrial and commercial equilibrium in a most lamentable manner. It would cause infinitely more misery in every respect than the abuse of intoxicants could produce within centuries. Hence, the advocates of prohibition have absolutely no economic basis for their claim.

Are their efforts justifiable from a moral point of view?—Let us see.

According to an oft-quoted Roman saying, abuse is not an argument against proper use.. If the contrary were true, we would be wrong, not only in indulging in all those enjoyments which distinguish man from brutes, but we would also be wrong in exercising any human virtue. Abuse means carrying to hurtful excess that which is beneficent when moderately used, exercised or indulged. Carried to excess, generosity becomes prodigality, frugality becomes parsimony, love becomes infatuation, self-respect degenerates into egotism, and so every human virtue, when carried to excess, becomes a vice, fraught with untold dangers to its victim. When abused, everything we call good is perverted into evil; and it is a rule as old as mankind, that abuse is under certain circumstances as inseparable from use, as effect is from cause. If, therefore, the prohibitionists' view, that abuse *is* an argument against proper use, could be put into effective practice in everything that is liable to be, and is being, abused, man would sink to the level of brutes. Even where the evils of the abuse over-balance the advantages of proper use, there can be no justification, from a moral point of view, for legislation against the latter, instead of the former; how much less then, in a case like ours, where the proper use is, and always has been, a blessing to millions, while only a few thousands suffer from abuse.

Intoxicants have been civilizers of nations. Wine civilized ancient Greece, and no intelligent man need be told of the beneficial effects of the cult of Dionysos, the rapture-bringer, on the intellectual development of the Greeks; nor of the grand works of art and poesy we owe to that epoch of Greek culture in which the Dithyrambos was originated. Tragedy and Comedy, says Stoll, in "Gods and Heroes," date their origin from the festivals of the god of wine. The justly celebrated historian Gervinus saw an intimate connection between human progress and the development of vine-culture. Wine created social bonds and social forms, and in so much as the intellectual development of man depended on these social forms, in just so much wine must be accounted a civilizer. The use of intoxicants assumed the form of religious usages with many ancient nations. The Egyptians venerated their god Osiris as the inventor of beer, and their libations had an ethical significance. The beer of the old Germans played a prominent part in the religion and the ethics of the people. The German drank to his God; he profferred the cup to the friend as a pledge of his loyalty, to the stranger as a guarantee of the inviolability of hospitality. All his actions were given a deeper significance, a moral meaning and a binding force through the agency of drink. In more recent times the drinking customs lost some of their meaning, but their influence remained the same.—All the festivals which grew out of the use of intoxicants had an elevating, a refining and ennobling effect on the community. In the many ale-festivals of Old England, the social development of the islanders is reflected. The lamb-ales, leet-ales, midsummer-ales, Whitsun-ales, Scotch-ales, &c., all had great influence on social life.*

The opinion that the use of intoxicants and our intellectual and moral development are closely connected, is held by all thinkers. Recently, an English writer, not an Anacreonite, but a sober physician, wrote a pamphlet on this subject. In the April number of the Popular Science Monthly, we find the following paragraph:

"Dr. William Sharpe seeks to demonstrate that alcohol is a factor in human progress. Looking into the history of the subject, he finds that the vine and the product of the vine have been in olden times more intimately associated with man's intellectual growth and development than with his purely physical wants. The stimulus of alcohol, when judiciously controlled, 'always leads to active and higher mental efforts on the part of individuals,' thus producing a contrary effect to that of other stimulants, which tend rather 'to bring about a contented state of dreamy inaction' and to repress effort. 'To understand fully,' he says, 'the beneficial action of alcohol as regards mental development, we must first get a clear view of the value of those states of cerebral excitement which most people, though in varying degrees, experience something of, rising as they then do mentally above the level of what may be called their ordinary every-day thoughts. This is not difficult, if we bear in remembrance that it is during such periods of high mental activity, in which the mind, transcending the more circumscribed limits of reason, sweeps intuitively into the veiled and distant regions of universal truth, that all great conceptions arise and have arisen in times past, crudely at first it may be, but which, nevertheless, when reduced to order and embodied in works, have been of inestimable value to mankind . . . . The stimulus produced by alcoholic liquors, if not nearly of so high an order, is more easily called into play, while in a practical sense, the latent ability being present, it is more vigorous and effective as regards actual work. Hence the value of alcohol, as a stimulant, lies in the fact that it produces artificially and sustains temporarily that state of mental excitement or exaltation necessary to the conception and projection, though not to the detailed elaboration, of those enduring works that, whether in the domains of art, architecture, or engineering, are remarkable for boldness of execution, originality, and grandeur of design; and further, that it is the only manageable stimulant which, when used in moderation, and in the form of wine or

* We quote from Vogel's "On Beer" based on Grässe's Bierstudien, and Brand. Pop. Antiquities. For the celebration of Whitsun ale it was necessary to elect a lord and lady of the ale, who dressed as fantastically as possible for their office. The locality for celebrating the festivity was generally a long barn, where seats were arranged for the company. Then arrived the lads and lasses of the village for feasting and dancing, and the young men offered ribbons and other finery to their sweethearts. A clown and music enlivened the Company. * * On the day of Lamb-ale celebration a fattened lamb was let loose and the girls of the village, with their hands tied together, had to run after it, and she who caught it with her teeth was called Lady of the Lamb. The lamb was then served on the village green, and the day was given up to pleasure and merriment.

spirits, is not only not injurious, but conduces to the general health, while it favors both mental and physical development.' Dr. Sharpe also assigns to alcohol a beneficial agency in stimulating genial thoughts and feelings."

Dr. Everts, in his "What shall we do for the drunkard," propounds this query:

"Is it not, indeed, probable that were all brain stimulants, other than ordinary foods common to man and other animals, at once and forever annihilated, or the alcoholic varieties alone withheld forever from common use, that the result would be, in the course of time, deleterious to mankind, by reason of brain deterioration resulting from a loss of such food, and a consequent gradual (no matter how slow) return of the races to a more common level, at the expense of those who have accomplished the greatest departure upward therefrom?"

Henry Ward Beecher recently expressed the opinion that "the more brains men may have, and the more brain-work, the more they are apt to be addicted to some form of stimulants, milder or severer, and only now and then can you find a man who is absolutely simple in his habits, drinking water and eating bread and meat or vegetables. Nor among them do we find the most robust, the most absolutely industrious, the most persistently accomplishing specimens of men."

The moderate use of stimulants is, indeed, absolutely necessary to the well-being of three-fourths of the male inhabitants of nearly all civilized countries, and to do away with such stimulants would involve great intellectual loss to the community, and a moral deterioration of society for which the salvation of a handful of drunkards would no more be an offset, than the saving of the cost of pauperism, insanity and crime (caused by intemperance), would be an offset to the loss of revenues derived from the use of intoxicants, and the incalculable losses which would be entailed upon agriculture, industry and commerce by a destruction of the liquor-traffic.

The position of prohibitionists is not, then, founded on a sound basis, either economically or morally.—Now, let us examine what restrictive laws accomplish, and what the law properly should and can effect.

LIQUOR LAWS.

PROHIBITION.

"As long as he on earth shall live,
So long I make no prohibition;
While man's desires and passions stir,
He cannot choose but err."—(*Goethe's Faust.*)

Prohibition was first tested in the Garden of Eden—and failed. The fall of man and his free agency were the results. All the imperfections of our moral nature are, according to the Scripture, consequences of this first failure of prohibition; for had not Eve plucked and eaten the forbidden fruit, man would be perfect. As it is, we are foredoomed to sin and suffer for sinning, but we are free agents.

The term prohibition is not, of course, used here in connection with drink, although many eminent writers would have us believe that the forbidden fruit was of an inebriating quality.* In principle there is no difference between "Thou shalt not eat this" and "Thou shalt not drink that." It is sufficiently significant that, taking a biblical view of the matter—and that is the view which such good Christians as our prohibitionists are, should take—all human misery began with the first failure of man to submit to prohibition. And it is still more significant that the man whom the Creator selected as the propagator of his species after the deluge, first exhibited the defects of his moral nature by drinking to intoxication.

Prohibition and its failure are, then, as old as mankind. Intemperance, and the laws against it are of nearly equal antiquity.

That intemperance must have prevailed to a great extent among the ancient Hebrews is sufficiently clear from the story the Bible tells us of Lot and others, and from the fact that Moses thought it necessary to promulgate restrictive laws against inebriety.

The Egyptians were strongly addicted to the use of wine and beer, and intemperance was common with both sexes. At the time

* "Milton seems to have entertained the opinion that the fruit of which our first parents had eaten

——'Whose mortal taste
Brought death into the world, and all our woe,'

was of an intoxicating nature.

The Jewish doctors were of the same belief, and Dr. Lightfoot and many eminent theologians were impressed with the like opinion."—*Morewood's History of Inebriating Drinks.*

of the Pharaos, laws were enacted against drinking excesses, and it was then the custom to place a skeleton and funeral draperies upon the festive board, whenever the revelers threatened to transgress the ordinary limits of hilarity.*

In the year 2200 B. C., the Chinese Emperor Yute banished the inventor of rice wine from his realm, and prohibited the use of that intoxicant, but without success. Grape wine, which was known in China as early as 1122 B. C., was also prohibited in subsequent centuries, partly for economic, partly for political reasons—the apprehension of a lack of cereals being at the bottom of the former, the fear of revolutions at the root of the latter. This prohibition, accompanied by the destruction of all vineyards, drove the Chinese people back to the use of the stronger rice wine and of opium.

Drunkenness was not unknown either in ancient Greece or in ancient Rome. Lycurgus imagined that he could curb the desires of his Spartans by exhibiting to them, on a fixed day of the year, a number of intoxicated islanders, who had been made to drink to excess by his order. His efforts seem to have been put forth in vain, however. In Athens, according to the laws of Draco and of Solon, death was the punishment for those who walked the public streets in a state of intoxication. Unlike our hyper-sentimentalists of temperance proclivities, who hold intoxication to be an excuse for crime, Pittacus of Mytilene caused a double measure of punishment to be inflicted for all crimes committed under the influence of intoxicants. Plato forbade the use of wine by minors under eighteen years of age, but granted all possible latitude to men of forty years of age, to whom he recommended frequent indulgence, encouraging them to abandon themselves to the joys of the banquet, to invite Bacchus to partake, and bring with him that divine liquor which he gave to man as a panacea with which to restore the vivacity of youth; sweeten the austerity of age, dispel its sorrows and mollify its harshness.

The drinking bouts of the ancient Romans excelled by far those of the Greeks. What must have been thought of drunkenness during the reign of Tiberius may be inferred from the fact that this emperor, surnamed Biberius (the bibler), appointed Pison Prefect of Rome for having passed two days and nights with him at the

* Geschichte des Weins u. der Trinkgelage, von Dr. R. Schultze. Intempérance et Misère, par J. Leffort. Historische Schriften, von G. G. Gervinus (Geschichte der Zechkunst), and Morewood's work.

drinking board, witnessing the feats of inglorious Novellius Torquatus, who was surnamed Tricongius from his ability to swallow three congii (about three quarts) of wine at one draught. Long before the reign of Tiberius sumptuary laws had been enacted, but they failed to check the evil.

The Gauls were no more distinguished for sobriety than their neighbors. A radical measure, not unlike that advocated by our prohibitionists, was carried out under Domitian (in the year 92), when that ruler ordered all the vineyards in Gaul to be destroyed. Beer then again took the place of wine.

The most striking illustration of the perniciousness of prohibition is that which the history of the Mahometans offers. The rigidly faithful observed the injunction of Mahomet with reference to wine, but their craving for a stimulant led them to the excessive use of opium—incomparably more destructive of moral and physical well-being than the strongest and worst liquors. While the faithful obey the prohibitory law from religious motives, the unbelievers ignore it, and resort to innumerable devices by which to evade the interdictory decree; and those who drink intoxicants must necessarily drink them solitarily and in secrecy. Morewood, in reviewing the secret drinking habits of the followers of Mahomet, says: "Where the influence of Mohametans has rendered the use of intoxicating liquors objectionable and penal, this prohibition has tended to render men artful and hypocritical. Although abstinence from inebriation is at all times commendable, yet, when carried to a complete deprivation, it has a contrary effect."

The Germans were hard drinkers at all times. The first glimpse history affords us of them reveals continuous drinking bouts. Nor is there a lack of laws against intemperance with them. The first restrictive liquor law is probably that of the Suevi, directed against the importation of wine. After vine-culture had been introduced by the Roman legions into the Rhinelands (281), intemperance grew apace; but no laws seem to have been enacted against it until the reign of Charlemagne. The capitularies of this great warrior and equally great law-maker abound in liquor laws, some of them showing very pointedly to what extent drunkenness prevailed at that time. Thus, one provision reads: "No Earl shall hold court unless he be sober;" from which it must be inferred that it was the custom of these judges to hold court while under the influence of intoxicants. The penal measures seem at first to have been extremely lenient. "Whosoever," one provision reads, "is found drunk in camp shall be compelled to drink water only until he shall have acknowledged that he has done wrong." Afterwards excommunication was added to the list of penalties, and when even this penal measure failed, castigation was resorted to, with no better success. It is of interest to know that Charlemagne, in his *Capitulare de Villis*, prescribed, that no people should be employed who did not know how to brew beer. He seems to have thought, then, that his laws against intemperance would have been more efficacious, if his subjects would drink beer exclusively. The Council of Aix-la-Chapelle (817) attempted to carry out a sort of Gothenburg system, in trying to regulate the consumption of wine and beer in each community. Fruitless endeavors, one and all! Not one of the laws and regulations, whether lenient or harsh in their penal provisions, accomplished the desired result.

With the introduction of distilled liquors, which were at first regarded and used as "death-preventing" medicine (hence the French name, eau-de-vie), inebriety assumed greater proportions and a new character. The use of these intoxicants first became general at the close of the fifteenth century in Hungary; subsequently in Bohemia, Poland, Pommerania and Russia. In the year 1581 the English soldiers, engaged in war in the Netherlands, were furnished brandy, and soon used it to excess. It was about that time that the first temperance society, excepting one, was organized by Maurice of Nassau, under the title of "Order of Temperance," to which noblemen only were admitted. The members pledged their word to drink *no more* than seven beakers of wine at each repast. They were, however, allowed to drink beer as much as they liked. Ardent spirits were prohibited, the penalty for drinking one glass of brandy being a deduction of two glasses from the regular allowance of wine for each offence. In 1524, the Margrave of Hesse prohibited the use of distilled liquors, and laws of like purport were enacted in Saxony, Würtenberg and Brandenburg.

Drunkenness existed in Britain when the Romans invaded that island. Under Saxon rule the evil experienced no diminution. With their political and personal freedom—the foundation of England's present state of civic liberty—the Saxons also transplanted to the new soil their social and martial drinking habits. Ale or beer was their common drink. The kings of the West Saxons must have been "enormously" fond of these drinks, since they exacted an annual tax of twelve ambers of ale from every owner of ten hides of land. Mention has already been made of the law (958) for the suppression of all ale-houses, excepting one in every village. Previous to the enactment of this law, Edgar had decreed that all drinking vessels should be provided with pegs, and that no guest at an ale-house should be allowed, under legal penalty, to drink beyond the next peg as the beaker went the round of the table.

In subsequent periods we find that whenever wars, revolutions, misrule, the profligacy of the nobles, epidemics, or like calamities disturbed the country, intemperance grew apace, and legal methods were then devised to check excesses in drinking, just as if these had not been merely one of many symptoms of the disease of the civic body. Thus, to cite but one instance, after the struggle between the houses of York and Lancaster had ended, lawlessness prevailed in England to an alarming extent. This was, of course, the result of the dissolution of all social bonds, the insecurity of life and property, and the daily recurring deeds of atrocity which had attended the revolution. Intemperance is not apt to grow less, when all the brutal passions of man are inflamed by the clangor of arms; but it is only one effect, not the cause. The laws enacted after that revolutionary period seem, however, to have been dictated by the conviction, that intemperance was the cause of lawlessness. In 1552 an act was passed prescribing the binding of keepers of ale-houses and tippling-houses by recognizances, and in 1554 the number of such houses, as well as of taverns, allowed to be licensed, was prescribed by law, and the drinking of wine on such licensed premises was peremptorily forbidden. This law had an effect, diametrically opposite to its object, as lawlessness increased enormously.

Prohibition in some form or other was often resorted to in England in order to curb the tendency to excesses, but all such measures failed.

Enough has been said to prove that there is not an age, however far removed from our time, when intemperance was not complained of as an evil of more or less grave consequences, and to prove, furthermore, that restrictive laws, although as old nearly as mankind, have failed to attain their end.

There is not a sumptuary law proposed in modern times for which a prototype cannot be found in the world's history; not one measure, advocated either by prohibitionists or temperance apostles, that has not been put into execution before. And still intemperance prevails to-day, as it did when Noah got drunk. Among all such laws none have failed more signally than those of a prohibitory character, whether they related to the use of fruit, of drink, or of tobacco. They either proved totally abortive, or led to vices incomparably more pernicious than those which they were intended to eradicate. The severest punishments, even the death penalty, as we have seen, failed to deter man from following that mysterious instinct which rendered the Creator's prohibition a failure. The decrees of the most absolute despots of modern or ancient times, the rulers of Russia, did not stop the use of tobacco, although mutilation of the body, life-long imprisonment and even death were the punishments meted out to offenders. Prohibition only proved successful when it assumed the form of a religious tenet, as, for example, in the case of Mahomet's decree; and then it gave rise to evils compared with which drunkenness seems almost like a virtue.

The fundamental condition of the success of prohibition would be a complete metamorphosis of man's moral and physical constitution—a nullification of that decree of the Creator, by virtue of which man, after the first failure of prohibition, and only on account of it, was left in the position, which Goethe describes in the lines that form the motto of this chapter.

Prohibition pre-supposes an unnatural condition of man, and is, therefore, an impossibility. The history of the past demonstrates this; the events of the present bear it out. Notwithstanding all that has been said and written to the contrary, it is a matter of positive official record, that prohibition does not, in our time, prohibit in any sense.

If it did prohibit, would not the returns of the Internal Revenue Office exhibit blank spaces under the head of liquor taxes opposite the names of States in which prohibition is the law? One will look in vain for such evidences, either in Maine or Vermont, in Iowa or Kansas.

The state of the liquor traffic may change, fluctuating for better or worse, as the methods of the administrations of these States become more or less rigid, more or less lenient; but under no circumstances has it ever, in any instance, been abolished as the law ordains. The only sure and invariable result has been, and is, that the secret traffic becomes all the more obnoxious, all the more dangerous to the security and morality of society, in proportion as the mode of executing the law grows more tyrannical, more directly subversive of those principles of personal liberty over which every American citizen naturally watches with the greatest solicitude. In considering the showing of the returns of the Internal Revenue Office in this connection, it should be borne in mind, that aside from the sale of ardent spirits, legitimately carried on under license from the National Government, there is, as we shall presently prove, in every State, where a prohibitory law is in force, a very large traffic which pays neither local nor national duties. And there is nothing strange in this; the injustice of the local law frequently produces that spirit of defiance, which manifests itself at all hazards. Hence, significant as the showing of the following table is, it does not exhibit the full extent of the failure of prohibition.

Kansas.—The following table, comprising nearly all the counties of the State of Kansas, shows whether prohibition prohibits there. As will be seen from the letter of transmittal at the foot of this page, the information was originally furnished by villages and towns, nineteen of which could not be located by counties, so that in some instances not all the licenses in force are included. The nineteen villages or towns are: Bethany, Baker Diggings, Beaver Creek, Chatauqua Springs, Cantonment, Eagle Springs, Fainge, Fort Hays, Fort Supply, Fort Reno, Fort Sill, Fort Gibson, Honeyville, Lazette, Leonard, Mulberry Ranch, Newburg, Sand Creek Ranch and Warwick. In 1880 in these nineteen places there were issued, in the aggregate, five retail liquor dealers' stamps; in 1881, six; in 1882, sixteen; and in 1883, twenty.

UNITED STATES INTERNAL REVENUE
COLLECTOR'S OFFICE, DISTRICT OF KANSAS,
LEAVENWORTH, Nov. 23, 1883.

Dear Sir: I to-day send you by express, as you requested, the number of retail liquor dealers' stamps issued in the District of Kansas during the tax years 1880, 1881, 1882 and 1883 up to the present time. The tax year commences on the first day of May. I have them alphabetically arranged by towns, so that any one can be referred to in a moment.

Respectfully,

JNO. C. CARPENTER.

List of Retail Liquor Dealers' Stamps Issued in the District of Kansas in 1880, 1881, 1882 and 1883, Arranged by Counties; also the Vote For and Against the Prohibitory Amendment:

Counties.	Number of Retail Liquor Dealers' Licenses.				Vote on Prohibitory Amendment	
	1880	1881	1882	1883	For.	Against
Allen	13	14	17	17	1305	951
Anderson	14	13	21	18	909	870
Atchison	102	105	111	121	1343	3147
Barbour	12	4	9	10	220	213
Barton	12	24	30	15	490	1058
Bourbon	38	35	44	25	1410	1964
Brown	7	8	10	6	1345	1288
Butler	26	21	26	26	2211	1141
Chase	14	14	12	19	597	660
Chautauqua	14	9	21	21	1051	819
Cherokee	48	39	32	32	2421	1944
Cheyenne	0	0	2	1	0	0
Clark	0	1	3	1	0	0
Clay	11	22	22	21	1296	907
Cloud	29	34	37	35	1454	1261
Coffee	15	22	23	21	1025	1209
Comanche	0	0	0	0	0	0
Cowley	32	14	17	26	3243	870
Crawford	35	20	25	36	1655	1469
Davis	22	14	20	17	628	607
Decatur	5	3	3	2	146	251
Dickinson	22	25	27	33	1477	1222
Douglass	33	35	39	35	2711	1602
Doniphan	41	45	42	44	821	2150
Edwards	5	4	4	5	121	194
Ellis	15	13	15	14	355	463
Elk	23	17	19	22	1232	564
Ellsworth	8	6	11	12	611	781
Ford	24	31	34	37	125	488
Franklin	21	16	15	16	1967	1293
Gove	7	3	5	3	0	0
Graham	5	2	2	2	207	358
Gray	5	5	7	4	0	0
Greenwood	19	10	16	10	1059	941
Hamilton	5	7	7	7	0	0
Harper	8	7	12	26	424	316
Harvey	22	20	44	29	1148	858
Jackson	11	11	14	12	1056	1098
Jefferson	32	26	31	31	1306	1723
Jewell	21	12	15	13	1557	1256
Johnson	20	10	12	19	1545	1787
Kearney	2	4	6	0	0	0
Kingman	6	2	3	9	265	346
Labette	52	33	39	50	2082	2123
Leavenworth	162	150	195	202	1486	3882
Lincoln	10	3	3	8	613	733
Linn	11	24	23	28	1494	1292
Lyon	30	20	35	41	2337	877
Marion	14	13	22	13	1020	825
Marshall	42	49	58	50	1428	1853

Counties.	Number of Retail Liquor Dealers' Licenses.				Vote on Prohibitory Amendment	
	1880	1881	1882	1883	For.	Against
McPherson	34	13	17	14	2134	912
Meade	1	0	2	0	0	0
Miami	22	19	29	23	1488	1751
Mitchell	20	19	20	20	1348	1178
Montgomery	44	31	34	42	1939	1250
Morris	14	3	9	12	895	885
Nemaha	25	20	14	19	1213	1185
Neosho	20	13	23	31	1528	1164
Ness	3	0	0	0	200	216
Norton	5	2	2	3	575	491
Osage	42	61	59	55	2287	1684
Osborne	11	11	12	8	1035	873
Ottawa	12	10	13	12	1163	835
Pawnee	9	2	3	4	604	218
Phillips	17	13	14	15	978	708
Pottawatomie	39	36	39	31	1549	1475
Pratt	8	3	3	3	151	142
Rawlins	3	4	1	2	0	0
Reno	14	25	33	28	1006	932
Republic	24	23	24	24	1330	919
Rice	21	20	16	20	1087	625
Riley	24	14	22	21	1178	828
Rooks	10	6	5	9	503	696
Rush	3	2	4	5	315	305
Russell	6	9	8	7	443	655
Saline	23	17	19	18	1410	1207
Sedgwick	57	57	62	66	1868	1716
Sequoyah	2	3	2	1	0	0
Shawnee	57	76	127	101	3159	2513
Sheridan	1	0	0	0	101	69
Smith	9	5	6	7	1274	851
Stafford	3	1	1	3	393	301
Sumner	82	51	59	59	2394	1201
Trego	6	5	2	5	220	120
Wabaunsee	10	9	10	9	622	990
Wallace	4	2	1	1	0	0
Washington	39	37	42	48	1112	1610
Wilson	25	19	20	19	1487	1069
Woodson	9	5	9	10	748	530
Wyandotte	71	115	141	142	1222	2481

Whatever may be said of the force of popular sentiment manifested in the matter of prohibition in the State of Kansas, it is pretty evident from the foregoing table, that this power did not tend to diminish the number of retail liquor dealers' stamps issued in the State; and not the least singular phenomenon is the increase of this class of licenses in many of the counties where the majority in favor of prohibition was more than commonly large. In Allen County 1,305 votes were given for the amendment, and only 951 against it, yet the number of licenses increased from 13, in 1880, to 17, in 1883;

in spite of 1,296 votes cast for prohibition, against 907 contrary votes, in Clay County, the number of licenses rose from 11 to 21 within three years; the abstinents of Butler polled twice as many votes for the amendment as their bibulous opponents did against it, yet the number of saloons remained stationary in 1880, 1882 and 1883.

It is true that popular sentiment, viewed in the light of election results, experienced a very marked revulsion within two years after the adoption of the prohibitory amendment; but what does this prove, if not that the sentiment for prohibition could not have been very strong in 1880? The vote on the amendment stood 92,302 for, and 84,304 against it; the majority in favor of it being 7,998. In the election of 1882 the champion of prohibition, Mr. St. John, received only 75,158 votes. The two other candidates, Hon. George W. Glick and Mr. Robinson, respectively received 83,237 and 20,933 votes.—In the strongholds of prohibition the mutation of feeling was not so striking, as far as this election result is concerned, but there is absolutely no possibility of contradicting the showing of our comparative revenue table, to the effect that popular sentiment, even where it seemed strongest according to number of votes cast, proved utterly powerless to prevent the augmentation of the liquor traffic, much less to terminate the sale of intoxicants, as the law ordains.

In this case prohibition not only fails to prohibit, but it is the direct cause of immorality and a prolific source of social hypocrisy and political degradation. The only thing it has accomplished in the way of suppressing intoxicants, is the banishment of fermented beverages, the very thing which, as the sum of experiences of all ages teaches us, should by all means be averted.

In many minor communities the law may be strictly enforced, because the majority of citizens are conscientious abstainers; but the neighboring communities, living under the same prohibitory law, allow saloons to multiply as fast as the demand for drink makes it necessary, so that, in the aggregate, the number of drinking places has increased, instead of being wiped out entirely.

In its operation, the law practically amounts to local option, but without the redeeming features of this system. It accomplishes its objects in a few isolated cases, but is absolutely nugatory in the majority of instances. It does not, happily, stop the use of stimulants, but it unfortunately encourages the abuse of them; it corrupts the drinker, whom no human or Divine law can deter from indulging his appetite, and creates contempt for the law in all.

Popular sentiment must have been very much at fault then in this case, or it was not rightly understood from the beginning. If neither is the case, why is the law practically a dead letter? De Toqueville's confidence in the vitality of our institutions was founded upon the knowledge of the power of popular sentiment to create laws and maintain them, on the one hand, and to completely nullify them without violence, on the other. In his "Democracy in America" he says that American legislators depend almost entirely upou the intelligence of the citizens, leaving it to the personal interest of all to live according to law. Such flagrant violations of the law as we see perpetrated in Kansas would, it seems then, be impossible if popular sentiment had sustained prohibition from the beginning.

It might be urged that the result of the vote on the amendment showed popular feeling to have been in favor of prohibition; but that would be a misstatement. The result only reflects the antagonism between two political parties—an antagonism that was utilized by a handful of well-organized, energetic theorists in furtherance of a measure to which the majority of voters on their own side were perfectly indifferent. No one pretends to say, that more than one-fourth of the 92,302 citizens, who voted for prohibition in Kansas, are total abstainers. Then why did the remaining three-fourths vote for it? Simply because to them it was a party measure which, if carried out, would not prevent them from indulging their appetites in the privacy of their dwellings.

But a man who votes for prohibition without being himself an abstainer makes himself guilty of a falsehood, just as he who votes for it from conviction is guilty of a tyrannical purpose, as far as the personal liberty of his neighbor is concerned. Thus, falsehood and tyranny are the parents of prohibition; and is it reasonable to expect that such an ill-begotten thing should thrive? At all events, it does not do so in Kansas, where the state of affairs of to-day is an exact counterpart of the situation in which Michigan had been placed by prohibition up to 1875.

Michigan.—The effects of prohibitory laws have nowhere been more graphically illustrated than in this State, in which prohibition was the law for twenty years. A prohibitory clause was inserted in the constitution of that State in 1850; and in 1853 the "Maine Liquor

Law" (slightly amended in 1855) was enacted and remained in force up to the year 1875. The evils that grew up under this law defy description. In 1874 there were over *six thousand places* in the State where ardent spirits were sold, and all of them were conducted openly. What the law had proscribed, ultimately became more powerful than the law and its executors, and public feeling abetted the law-breakers. There was, indeed, one powerful example of "laws outlawed by necessity."

As soon as the law had fallen into public contempt, no limits could be fixed for the audacity of the law-breakers, and the only remedy at hand was the revocation of the prohibitory clause, and the law based upon it. The best citizens, and among them nearly every sincere advocate of temperance, united in a grand movement against prohibition, and the result was a brilliant vindication of outraged common-sense.

It cannot be said that the law was insufficient. "The prohibition," said Hon. George W. Moore, before the Detroit Board of Trade, some time ago, "was as absolute as it could be made. The ingenuity of the ablest lawyers, preachers, business men, legislators and women, was exhaused in devising penalties and means of enforcing them. Liquors were declared no consideration for a debt, and any sale of other goods where liquors were part of the trade, was declared unlawful and the debt could not be collected; that every person injured by such sales should be able to sue the seller and recover damages; that owners of the buildings should be also liable; that any lease of premises where liquor was sold could be declared forfeited; that each act of selling should be a separate offense, punishable with fines not exceeding one hundred dollars and imprisonment up to six months, until the liability of every liquor dealer in the State would aggregate perhaps hundreds of thousands of dollars, and imprisonment for many lifetimes. Common law rules of evidence were changed to make convictions easier, and the simple solicitation of any intemperate person to drink subjected the inviter to the penalties provided for the seller."

All to no purpose! The evil did not cease until its source was destroyed; and then, and not until then, did the people of Michigan realize what an awful burden they had borne for twenty long years. Under the new law, placing the liquor traffic under rational excise restrictions, two thousand of the lowest groggeries were swept away during two years; and it was thus clearly demonstrated that prohibition, far from stifling the craving for strong drink and destroying the opportunities for gratifying that craving, calls forth far more drinking places than the wants of reasonably regulated communities would justify under ordinary circumstances. Offenses against public peace and order decreased in an uncommon degree; the liquor interest was made to bear its share of the burden of taxes, which had erstwhile been borne wholly by other interests, and the brewing industry received a new impetus from the rapidly developing refinement of the drinking habits of the people.

It will presently be shown that nearly all advantages that had grown out of an equitable and just management of the liquor traffic are now being neutralized, if not entirely paralyzed, by the system of high licenses.

Massachusetts.—Prohibition failed as ignominiously in this State as everywhere else. To describe the course of its growth and failure would be but a repetition of what is stated under the head of Kansas and Michigan, if it were not for one very significant feature of the operation of this pernicious law in one of the most enlightened commonwealths of our land. There it was proven that prohibitory laws are not only tyrannical in principle, but that, to be anything more than mere farces, their execution requires measures that must be repugnant in the last degree to every sincere lover of liberty. The report of the Joint Special Committee of the Legislature of Massachusetts on the subject of a License Law, rendered on the fourteenth day of May, 1867, contains the following passage, full of the utmost importance to every patriot:

"In our republican form of government, we have always recognized the fact that no criminal laws can be faithfully executed (and therefore should not be enacted) which are not sustained by the moral convictions of the people. When we make changes in them from time to time, we are content to leave the execution of the new laws with the ordinary instrumentalities. For the administration of our entire criminal code, old laws and new laws, we have relied upon the vigilance of ordinary municipal officers to complain of violations; the fidelity of prosecuting officers, elected by the people, to take charge of the complaints or indictments when made or found; the honor and good sense of juries, selected under long-established and well-known rules, to convict or acquit, according to the law and the evidence, and the discretion of the judges, in case of conviction, to impose reasonable sentences. All these regular and ordinary methods were open for the execution of the statutes upon the sale of liquor. If the moral judgment of the people approved the law, there was no sufficient reason in the nature of things why police officers, district attorneys, juries and judges should not be as prompt and decided in doing their respective duties by this as well as other laws. Yet the course of the supporters of the present statutes seems to indicate great distrust upon their part of all these parties, or rather that there is something in the law so different from the principles of our ordinary criminal legislation, and so repugnant to the popular instincts, that new and arbitrary methods are necessary to enforce it.

Every city and large town has its local police, which had been found effective enough in preserving the peace, and prosecuting violations of State and municipal laws. Yet the execution of this law could not, it was thought, be safely entrusted to them because they were not sufficiently eager to prosecute ; and hence *a system of State constabulary was adopted, until that time unknown in this country and in other republics, and borrowed from monarchial countries.*"

But even so, with all this machinery of tyranny, the law remained a dead letter. This legislative committee did its work very thoroughly and impartially, hearing both sides with equal patience and intelligent attention, and conducting its inquiries into the effects of prohibition in the broadest possible, yet most thorough manner. The conclusions reached were, that the law was unjust, illogical in theory, and nugatory in practice; that it did not prohibit, but that it did transform very many dwellings into secret rum-shops; that it corrupted private and public morals, increased crime and pauperism, and consequently augmented the burdens of taxation.*

Maine.—The prohibitory law is still nominally in force in this State, but in fact it is inoperative. Even Mr. Dow has not the temerity to assert that prohibition prohibits in his State. All he dares to claim for his pet system is, that it has prevented the further growth of the evils of intemperance, and has transformed "Maine from the poorest State in the Union into one of the most prosperous."†

Prosperous? Let us see. While the population of the United States has increased at the rate of thirty-three per cent., Maine's population, after retrograding during one decade and remaining stationary during another, finally crept upward at a snail's pace—three

* See Appendix of said report, pp. 238, 241, 314, 336, 339, &c. The law was repealed in 1868 and license substituted. In 1869 it was re-enacted, with the exclusion of cider. In 1870 it was amended so as to permit the sale of malt liquors in places in which the citizens did not prohibit such sale. In 1871 the sale of malt liquors was made dependent on a vote in favor of it. In 1873 prohibition pure and simple was decreed by the legislature, and in 1875 the license system again went into force and effect.

† Neal Dow's answer to the report of the English Consul at Portland, Maine.—*London Times of the 6th of October, 1883.*

per cent. being the climax of her progress. Progress, indeed! Look at these figures:

	Population in 1880.	Number of Paupers in 1880.
Iowa	1,624,615	2,133
Indiana	1,978,301	3,965
Michigan	1,636,937	2,300
Minnesota	780,773	496
New Jersey	1,131,116	2,981
Maine	648,936	3,211

With a smaller population than Minnesota, Maine has over six times as many paupers as that State. With a population larger by almost a million than that of Maine, Iowa has only 2,133 paupers, or 1,078 less than Maine. Michigan is ahead of Maine in point of population by nearly one million, but the latter State takes the palm in point of pauperism by 911.

Strange indications of prosperity, these!

Mr. Dow derives comfort from the thought, that Maine is not to-day what he thinks it would have been, if prohibition had not diminished the consumption of distilled spirits. He says, in the rejoinder referred to: "Our share of the national drink-bill would be now about $13,000,000, but $1,000,000 will cover the cost of all liquors smuggled into the State and sold in violation of the law." That is to say, that if Maine's population had up to this day remained in the crude moral and intellectual condition which prevailed at the time when the State's great industry, as Mr. Dow says, was the lumber trade, and when prohibition was introduced, the consumption of distilled spirits would cost Maine $13,000,000 annually. Mr. Dow seems to ignore the fact that the lumber trade was the cause of the coarse drinking habits of the people in Maine, just as it was the direct incentive to that effort of Dr. B. J. Clark, to which the organization of the first American temperance society, in Saratoga County, is to be ascribed. The coarse habits of the lumber-men; their rough out-door life, which denied them all the thousand comforts that the poorest laborer enjoys in a civilized community, led to those excessive drinking habits which alarmed Dr. Clark, as early as 1808. In Armstrong's History of the Temperance Reformation, we read:

"Alarmed at the prevailing custom of the region of country around him, teeming with lumber in all the towns and counties in the vicinity of the ever-rolling Hudson, in all which intoxicating liquors of variety and plenty were considered as commodities of necessity for the daily use and comfort of all, or almost every family, and *indispensable* for the treatment of friends in social life—alarmed, we say, at the prevalence and results of such a custom, after having projected the plan of a temperance organization, the doctor determined on a visit to his minister," etc.

Prohibition could do nothing for "all the towns and counties in the vicinity of the ever-rolling Hudson" (the prohibitory law of 1855 having been declared unconstitutional in 1856), yet Mr. Dow's brilliant argument applies to this fertile region quite as well as to Maine. In fact, if the people of the United States had not progressed in any direction—and that is Mr. Dow's basis of argument as to Maine—the per capita consumption of distilled spirits would to-day be anywhere from 11 to 15 quarts, instead of $4\frac{1}{3}$. It is not a question of what would have been, but of what is; and on this point Mr. Dow left no doubt on the mind of the readers of his reply in the *London Times*. He admits, in fact, that prohibition does not prohibit. It has, it is true, wiped out distilleries and breweries, but in the place of these legitimate industries, everywhere yielding large revenues to the State, it has called forth a regular system of smuggling, by which Maine is supplied with vastly more liquor than would be consumed under an equitable license law.

According to the testimony of impartial observers, all the pernicious results of prohibition are in full bloom in Maine. Fermented beverages, particularly malt drinks, are little used, but distilled liquors of the worst quality find ready sale everywhere; and if we were inclined to turn the tables on our opponents, we might say, that this accounts for the enormous increase of pauperism and insanity in Maine. It is the curse of all such unreasonably restrictive measures, that they injuriously affect the condition of the laboring people, whom they deprive of wholesome malt beverages and drive to the use of such ardent spirits as can be had for little money. The rich and the well-to-do are not in any way inconvenienced by such laws, and it is probably for this reason that they preserve a degree of indifference to prohibition, that would otherwise seem inexplicable. In this connection Mr. Dow made a very strange confession when he said, in his *Times* letter, that Consul Bird could know nothing of the working of the law because "his associations here (in Portland) are with most respectable people, none of whom have any sympathy with the temperance movement, much less with the policy of prohibition." Doubtless, these respectable people would manifest not only a lack of sympathy for, but a very energetic antipathy to, prohibition, if that obnoxious law infringed upon their personal rights as it does upon those of the poor laborer, whom it compels to become the unprotected customer of the proprietors of low dives. The compiler is informed by a responsible officer of one of the largest cities in Maine, that *the stanchest supporters of prohibition are contrabandists, whose lucrative trade would be destroyed if the State adopted an excise system worthy of a civilized commonwealth.* The same officer is of the opinion, frequently expressed by independent journalists, that if popular sentiment could ever be fairly and squarely tested, in reference to prohibition, the law would be smothered under a mountain of ballots. The question in itself and on its merits has never been voted on; it was always a side-issue in the struggle of political parties for power. If one would take the pains to read the newspapers and magazines of the time when the Maine law was in its infancy, he would find ample proof of this. Here is an excerpt from an article published in 1859, under the title "History of the Struggle in Maine:"

"It must suffice for our present purpose to recall to remembrance the two great parties into which American politicians are divided—viz., Republican and Democrat. The Whigs, and after a meteoric course of success, the Knownothings, though still a numerous body, may be disregarded in the consideration of the circumstances we are about to describe. The distinctions between the Repub. licans and the Democrats are radical. At the last presidential election, our readers will remember, Colonel Fremont represented one of these parties, while Mr. Buchanan was the nominee of the other. Irrespective of general policy, on the great American difficulty, the two parties hold opposite views. The Republican party is anti-slavery; the Democratic sympathizes with the feelings of the Southern States. As may be supposed, the majority of temperance men belong to the Republican party—*slavery and rum are too intimately associated to be dissevered in politics.** Rum figures largely in the slave traffic as a 'medium of exchange,' and avenges outraged humanity by binding the 'superior race' in a still more abject bondage than that of the chain and the lash. *Hence, it being rarely possible to present to the people for their vote a temperance issue uncomplicated with other party considerations, the temperance party in Maine and in other States has generally shared in the vicissitudes of the Republican party.* Of course, all Republicans are not temperance men; some are purely and selfishly politicians; and there have been instances in which, having used their votes, the Republican politicians have looked indifferently on the claims of the friends of sobriety."

* This brings to mind the fact that the outbreak of the "irrepressible conflict" put a stop to the prohibitory movement in spite of the "intimate connection" between "slavery and rum." Why was it thus? Because the party which fought slavery and rum (rum, of course, means all intoxicants) found a most powerful ally against slavery in the German element, which German element was, and is, just as earnestly opposed to prohibition as it was to slavery. It was this aid from so unexpected a quarter that obstructed the prohibitory movement for over twenty years. By the year 1857 prohibition had been voted on, and either adopted and enforced, or adopted and declared unconstitutional, in Maine, Delaware, Rhode Island, Massachusetts, Vermont, Michigan, Connecticut, Indiana, Iowa, New York, New Hampshire and Illinois. In all of the States named, in which the German votes were worth having, the prohibitory movement came to a halt as soon as the "late unpleasantness" began.

How forcibly these words remind one of the part prohibition plays in the political campaigns of our day! A few zealots, at the head of a column of blind, but well-meaning followers, invade the political arena and dictate terms to the party leaders, who are but too willing to promise anything and everything, with or without an intention of keeping their promises. If they keep them, it is not always because they hold the thing promised to be good, wise and necessary in itself; if they break them, it is because by doing so they hope to gain more votes, than by redeeming their pledges.

Popular sentiment had as little to do with the creation of the prohibitory law in Maine as in Kansas; and that is one, but only one, of the reasons of its failure there as elsewhere.

What is true of Kansas and Maine, of Michigan and Massachusetts, is also true of Vermont and New Hampshire, and of all those other States in which prohibition exists in the form of local option. This latter method appears to be in great favor with the Democrats of the South. In some instances the sale of intoxicating liquors within the limits of counties is directly prohibited by legislative acts; in others the legislature empowers the voters of certain counties to vote upon the question, and again in others a general law imparts that authority. In the State of Alabama the sale of liquor is prohibited either in parts, or the whole, of fifty-five counties. The State has sixty-seven counties. In Georgia, where the ordinaries of counties have power to grant or refuse licenses, local option prevails by the side of prohibitory laws, and high license acts. The prohibitory laws relate either to entire counties or to parts of them, and so do the laws fixing the license fees. Thus, for instance, by an act passed on the 26th of September, 1879, the sale of liquors was prohibited within the limits of Irwin County. On the 2d of October, 1879, an act was approved, by virtue of which the license fee for selling liquors in the counties of Wayne, Liberty, Coffee and Appling, was fixed at $1,000.—At the end of last year prohibition was in force, either wholly or partially, in ninety-one counties of Georgia. In Maryland, where a fair license law is in force, there are about twelve or thirteen counties whose "qualified voters" have, by legislative enactment, been "enabled to determine by ballot whether intoxicating liquors or alcoholic bitters shall be sold." All these laws are nearly of the same tenor, except in so far as fermented liquors are sometimes included in, sometimes omitted from, the list of forbidden drinks. In these counties, according to reliable newspaper reports, a host of itinerant whiskey-sellers do a profitable business. Local option is, in fact, just as much a failure as prohibition. It does not accomplish its immediate object, and usually brings about the very reverse of what should be the ultimate object of every law, *i. e.*, the improvement of the moral and material condition of those who live under it. It will always be secretly evaded* or openly and defiantly violated, and will in all cases retard that refinement of the drinking habits of the people from which alone genuine temperance can reasonably be expected; since excessive restrictions, indiscriminately placed upon the sale of *all* intoxicating liquors, have a tendency, as we have seen, to put malt liquors beyond the reach of the majority of drinkers, and to increase the quantity and deteriorate the quality of ardent spirits consumed.

THE HIGH LICENSE SYSTEM.

The great reputation of Americans for inventiveness rests largely upon their mechanical and industrial achievements; in legal matters we have displayed less originality than in any other respect. We have copied copiously, but not always wisely, from English statute books; and, strange to say, we have done this in one instance, even after the worthlessness of our model had been fully established. England's experience with high licenses should be a warning to our law-makers; but, unfortunately, it is not. Indeed, the history of taxation in England affords many very instructive illustrations of the dangers that attend unwise excise legislation.

* The ingenuity with which such laws are evaded is well illustrated in the following telegraphic dispatch sent to the New York *Times* from Atlanta, Ga., under date of February 9, 1884: "The revival of the earthenware business in Georgia is one of the curious results of the local option movement. The high licenses at first adopted limited the sale of liquor to country towns, and the adoption of local option by several contiguous counties forced a good liquor trade upon the nearest market town where it was sold. Just before Christmas it was noticed by Southern Express officials that a great number of jugs were put into the freight directed to parties in temperance counties. From Griffin to Carrollton, for instance, there was a large traffic carried on in jugs, filled, of course, with whiskey. Stewart County is known as "wet," but all counties around are local option, so that Stewart has to bear the blame and expense of drunken freaks of half a dozen of her neighbors. Mr. Cullom, of Aiken County, S. C., filled an order within a month from Savannah for 10,000 jugs. He also disposed of 2,500 in Waynesboro, Ga. As these jugs are used for illicit purposes, they are never used more than once, thus keeping up the demand for new stock. The new business has attracted the attention of manufacturers, and agents are now in Swainsboro, Statesboro and other places establishing depots for the supply of jugs. Rates have been received from railroads, and whenever a depot can be established within one day's wagon drive of a temperance centre, it will furnish liquor to all who need it. There are jug factories in Washington and Clark Counties, Ga., and several in the northern part of South Carolina, all of which feel the improvement in business due to the cause mentioned. Temperance men have endeavored in several instances to find a remedy, but seem to have been unsuccessful. In one instance an attempt was made to enjoin the delivery of jugs by the Southern Express Company, but the effort fell through, as there was no authority upon which such action could be based.

There is not an English writer of standing, who would venture to deny that if the drinking habits of the English people are more intemperate to-day than they were two centuries ago, it is owing to the exorbitant taxes with which the brewing industry was burdened from the beginning of last century.

Beer was at one time the favorite beverage of the English, we are told. It would no doubt have retained its ascendency over all other liquors—just as it has in Bavaria—if the government had not placed it beyond the means of the people by exorbitant taxation. From and after the Cromwellian era, the taxes on beer rose rapidly to an almost incredible height. In 1659 the tax on beer amounted, in the aggregate—for England and Scotland—to £374,456, exceeding by far, as Vocke says,* the income from all other excise duties. The revolutionary origin of these taxes did not deter the parliaments of the restored monarchy to continue them at the Cromwellian rate of 2*s*. 6*d*. per barrel of beer, costing more than 6*s*.; and 6*d*. per barrel, costing less than 6*s*. Under the two last Stuarts the taxes remained unchanged. In the reign of William III. they were raised first, in 1689, by 9*d*. and 3*d*. respectively; in 1692 and 1693 again, each time by the same amounts; and in Queen Anne's reign 3*d*. and 1*d*. respectively; so that in 1710 the tax on every barrel of beer, costing more than 6*s*., was 5*s*., and on every barrel costing less, 1*s*. The duties amounted to 83 per cent. of the value of the product.

In addition to these excise duties, a tax of 6*d*. per bushel was levied on malt, and one of 1*d*. per pound on hops, so that the product and its ingredients were trebly taxed. As in Germany, in Sweden and other countries, so also in England, it was the custom of nearly all well-to-do people to brew their own beer, and this was made to contribute to the public exchequer through a tax of 5*s*. per head of every household so brewing.

Distilled spirits were taxed at the same time, but in nothing like the rate of duties on malt beverages. The tax, which originally amonnted to but 2*d*. per gallon, was raised, it is true, in the reigns of William III. and Queen Anne—periods distinguished for the insatiable necessities of the public exchequer—but the increase did not amount to more than 8*d*. per gallon in 1710. Distilled spirits were, consequently, exceedingly cheap, compared with the cost of

* Geschichte der Steuern des Britischen Reiches. Von W. Vocke; p. 383. (Leipzig, 1868.)

beer, and the people were forcibly driven to the use of gin. The "gin epidemic" was the result of this rapacious system of taxation. In London there was, in 1725, one spirit-shop to every seven houses. In 1728 "high licenses" were resorted to as a measure supposed to diminish intemperance; but the change was not productive of any practical good; and, besides, this law was soon (1732) revoked. It was at this point that the state of things assumed that aspect which Smollett so graphically describes in the work already referred to. This was the time when, as that author says, "signboards proclaimed: 'Here you may get drunk for a penny, dead drunk for twopence, and have clean straw for nothing.'" In 1735 the oft-quoted Gin Act was passed, fixing the price of a license for the sale of distilled spirits, in quantities of less than two gallons, at £50, and imposing an excise duty of 20*s.* per gallon. The law-making power, whose rapacity had artificially created an appetite for ardent spirits, foresaw that so stringent a measure could not but invite evasion or provoke open defiance; and to guard against either, large rewards were offered to informers. The penalty for violations of the law was extremely severe, and everything was done to detect and convict law-breakers.

What was the result? To use Smollett's words, "the people broke through all restraint, and illicit spirit-selling assumed gigantic proportions." The consumption of distilled spirits increased enormously, and, although 12,000 persons were convicted of violations of the law and severely punished, the government was powerless to restrain the evil even within that broad latitude which originally led to the high license measure. The torrent of popular dissatisfaction carried away all barriers. Secret evasion and open and defiant violation of this law—in many instances informers and officers were chased like wild beasts by infuriated mobs—produced contempt for all laws, and widespread immorality was the inevitable consequence.

The utter worthlessness of high licenses as a temperance measure became so obvious that the obnoxious law was revoked in 1742, to make room for a moderate excise law, through the operations of which it was hoped to do away, at least, with those evils that were not results of intemperance, but consequences of the general lawlessness. If this measure had been accompanied by the abolition of the malt tax, it might have tended to wean the people from the taste for ardent spirits. As it was, the government continued experimenting alternately with high licenses and low licenses; cheapening ardent spirits in the latter case to such a degree that their use could be indulged in, without breaking the law, at a smaller cost than that of beer; and in the former case provoking utter disregard of the law. But, in the meantime, the beer tax was also increased, instead of being decreased. In 1760 the malt tax was raised to 9d., the hop tax to 1½d. The excise duty on beer, in 1761, was 8s. per barrel, costing over 6s., and 4s. 9d. per barrel, costing less than 6s.; In 1803 the former tax was raised to 10s. With few intermissions, these duties retained their upward tendency until Canning proposed a temperance reform in the way of a reduction of taxes on malt liquors (1822), which was followed by the abolition of the beer excise in 1830.

The production of ardent spirits and malt liquors, during the periods in which both kinds of beverages labored under heavy burdens, proves beyond the possibility of a doubt what the high license system must inevitably bring about. The production of malt liquors decreased with every new tax imposed upon it, while the production of distilled spirits increased most rapidly at the very time when it was most heavily taxed. The reasons of this are obvious. The illicit sale of ardent spirits can be carried on without very great danger of detection, while the illicit sale of malt liquors is well-nigh impossible.

When Canning's measure went into effect the consumption of malt began to increase at once. From 1821 to 1830 it rose from twenty-one million bushels to twenty-seven million, and continued rising after the abolition of the excise duty on beer. As soon as a new duty was imposed on malt, as during the Crimean war, for instance, the consumption again decreased.

It is asserted, and with good reason too, that the laws favorable to the manufacture of malt beverages have not increased the consumption in anything like the proportion of increase in population. But how could anything else be expected? After the nation had for nearly 150 years been compelled—there is no other word for it—to drink gin; after those edifying festivals which derived their origin from the use of ale, and which tended in such a marked degree to elevate the masses, had been relegated to oblivion; after the drinking habits of the people had become so thoroughly revolutionized in every respect—how could it be expected that malt beverages should at once, within a couple of decades, resume their old place in the favor of the people? Besides, the taxes on the product indirectly remained comparatively high, even after the reformatory measures mentioned. The taxes on malt were always rather high; according to Vocke's calculation the duties on the ingredients of beer amounted, in 1857, to 50 per cent of the market value of the product. In short, with all the relaxation of duties on beer and its ingredients, distilled spirits remained cheaper than malt beverages, and it is that which partly accounts for the fact, that the increase in the consumption of the latter drinks is so small. Another reason is, that during the fiscal proscription of beer, the English people became habituated to the use of other stimulants, as tea and coffee. The number of barrels of beer brewed in 1857 was 17,984,773, in 1869 it was 24,542,664; the increase being 36.40 per cent. In 1883 the production amounted to 27,140,891 as against 27,870,526 barrels in the preceding year.

This proves that it is easy to legislate a nation into intemperance, but that it is an exceedingly difficult task to counteract the evils of such unwise legislation. In considering the question of high licenses, as the term is understood here, we have no need of inquiring into the present state of things in England. Certain it is, that the general use of beer in England ceased, as soon as malt beverages and their ingredients were heavily taxed, and that when, after a long period of pernicious experimenting, the sale of distilled liquors was placed under the restriction of exorbitantly high licenses—beer still being taxed as heavily as before—the consumption of ardent spirits increased enormously, while that of malt beverages sank to almost nothing.

This is what has to be considered, nothing else; and from whatever point of view it may be done, the conclusion is inevitable that no measure ever proposed or executed, prohibition always excepted, has such a pronounced tendency, as the system of high licenses, to aggravate the evils of intemperance by forcibly driving the people to the use of ardent liquors, not to speak of the violations of the law to which it entices, and the increase of drinking places which it produces.

Michigan.—Having tested prohibition, and abolished it after a thorough trial of twenty years' duration, this State adopted the license system, making a wise discrimination between fermented and distilled liquors. The law of 1875 fixed the price of a license for the sale of fermented drinks at $40, and the other at $150;

subsequently these sums were raised to $65 and $200, and still later to $200 and $300 respectively. The section of the law, fixing the latter rates reads:

SECTION 1. In all townships, cities, and villages of this State there shall be paid annually the following tax upon the business of manufacturing, selling, or keeping for sale, by all persons whose business, in whole or in part, consists in selling, or keeping for sale, or manufacturing distilled or malt liquors, or mixed liquors, as follows: Upon the business of selling or offering for sale spirituous or intoxicating liquors or mixed liquors by retail, or any mixture or compound, excepting proprietary patent medicines, which in whole or in part consists of spirituous or intoxicating liquors, the sum of three hundred dollars per annum; upon the business of selling or offering for sale by retail any malt, brewed or fermented liquors, two hundred dollars per annum; upon the business of selling brewed or malt liquors at wholesale, or at wholesale and retail, two hundred dollars per annum; upon the business of selling spirituous or intoxicating liquors at wholesale, or at wholesale and retail, five hundred dollars per annum; upon the business of manufacturing brewed or malt liquors for sale, if the quantity manufactured be less than fifteen hundred barrels, sixty-five dollars per annum, and twenty-five dollars upon each additional thousand barrels or part thereof; upon the business of manufacturing for sale spirituous or intoxicating liquors, five hundred dollars per annum. No person paying a tax on spirituous or intoxicating liquors under this act shall be liable to pay any tax on the sale of malt, brewed, or fermented liquors. No person paying a manufacturer's tax on brewed or malt liquors under this act shall be liable to pay a wholesale dealer's tax on the same.

The advocates of high licenses claim, contrary to what has been the experience in England, that this method reduces the number of saloons and consequently diminishes the opportunities for "getting drunk"; that it does away with the low dives, and increases the revenues. Let us see whether this claim is justified by the actual state of things in Michigan.

From the records in the office of the Auditor General of the State it appears that there were, in 1882, three thousand four hundred and forty-four licensed saloons, against three thousand nine hundred and seventy in the preceding year, there being a reduction in the number of saloons of five hundred and twenty-six. The revenues amounted to $550,185 in the former year, and to $913,684 in the latter. If the "high-license" law had for its object simply an increase of the revenues, it would undoubtedly have to be regarded as a complete success. But the fiscal consideration is said to be secondary only, the main object being of a moral nature, *i. e.*, the checking of intemperance. A reduction in the number of saloons does not in itself argue a decrease in drunkenness, unless it can be demonstrated that the consumption has correspondingly decreased.

Well, notwithstanding the report of the Auditor General, neither a reduction in the number of saloons, nor a decrease in consumption has taken place.

The number of saloons *licensed by local authorities* has no doubt been diminished; but illicit selling is carried on in a great number of saloons—illicit only so far as the evasion of the local, not the United States, revenue laws are concerned. One example is as good as a hundred. The following table shows the number and kind of licenses issued in the City of Detroit during 1882 and 1883:

Number of Licenses.	Amount of Tax.	Rate per Annum.	Kind of Business.
			1882.
15	$7,500 00	$500 00	Wholesale Spirituous Liquors.
331	99,300 00	300 00	Retail Spirituous Liquors.
208	41,600 00	200 00	Retail Malt Brewed or Fermented Liquors.
17	1,125 00	65 00	Brewer's License (1,500 Bbls. or less), $25 for every additional 1,000 Bbls.
104	15,483 43		Various kinds for fractional portion of year.
675	$165,008 43		
			1883.
11	$5,500 00	$500 00	Wholesale Spirituous Liquors.
209	62,700 00	300 00	Retail Spirituous Liquors.
305	61,000 00	200 00	Retail Malt Brewed or Fermented Liquors.
22	1,430 00	65 00	Brewer's License (1,500 Bbls. or less), $25 for every additional 1,000 Bbls.
156	19,913 55		Various Licenses for fractional portion of year.
703	$150,543 55		

The following extract from a letter of the U. S. Internal Revenue Collector at Detroit, Mr. J. H. Stone, sustains our assertion:

"I have caused an examination to be made of the special tax record of this district for the years ending April 30, 1883, and April 30, 1884, and find that there were issued for those years special tax stamps for retail liquor dealers in the city of Detroit as follows: 1882–3, 3,919, and 1883-4, 996."

Here we have a difference between United States and local licenses, in 1882–3, of two hundred and sixteen in Detroit alone;

hence there are two hundred and sixteen places in which intoxicants are sold unlawfully. That these violations of the law are the result of high licenses, cannot be doubted. In 1880 the number of places having local licenses was 905; in 1881 it was 812, and 675 in 1882. The number of United States revenue licenses issued during 1882-3 was only slightly larger than that of local licenses issued in 1880, so that it is clear that the local law did not affect the actual number of drinking-places, while it made two hundred and sixteen law breakers. Neither the quantity consumed nor the number of saloons were in any way affected. All the law accomplished was to entice those who were formerly law-abiding citizens into violations of the law.

The demoralizing effect of high licenses becomes still more obvious from the fact that the unreasonably high price of beer licenses must necessarily affect the consumption of malt liquors, and increase intemperance. No better proof of this can be adduced than the following table, showing the amounts received for licenses issued by the United States Revenue Office in the first district of Michigan, in which the city of Detroit is situated:

Year.	Retail Liquor Dealers @ $25.	Wholesale Liq'r Dealers @ $100	Brewers less $50.	Brewers more $100	Retail Malt Liq'r Dealers $20.	Wholesale Malt Liquor Dealers, $50.
1880...	$35,428 11	$1,725 00	$683 34	$3,083 33	$2,503 36	$1,145 84
1881...	37,516 47	2,050 00	733 34	2,950 00	1,979 19	1,050 00
1882...	33,905 46	2,212 50	575 00	3,300 00	989 18	947 50
1883...	37,873 64	2,237 50	450 00	3,091 67	758 35	760 41

This tells the whole story very forcibly. While there is, from 1880 to 1883, an appreciable increase in the amounts collected from retail and wholesale liquor dealers, there is a vast decrease in the amounts collected from retail and wholesale dealers in malt liquors. In 1880 the United States received $35,428.11 from retail liquor dealers; in 1883 the revenues from this source amounted to $37,823.64. In 1880 the revenues from malt liquors amounted to $2,503.36, in 1883 to $758.35.

To show that prohibition and high licenses have nearly the same effect, inasmuch as both favor the increase of places for the illicit sale of distilled spirits and the decrease of places where malt liquors are sold, we give the following table, for which we are indebted to Hon. Walter Evans, Commissioner of Internal Revenue:

Kind of Liquor Law.	Fiscal Years ended June 30.	Retail Liquor Dealers. Number.	Retail Dealers in Malt Liquors. Number.
Prohibition	1863	2,248	....
"	1864	2,218	....
"	1865	3,442	....
"	1866	4,087	....
"	1867	4,223	....
"	1868	4,604	....
"	1869	5,537	....
"	1870	5,020	....
"	1871	5,095	....
"	1872	5,846	....
"	1873	8,488	79
"	1874	6,392	122
Moderate License	1875	5,680	219
" "	1876	4,828	572
" "	1877	4,384	408
" "	1878	4,505	569
" "	1879	4,373	440
" "	1880	4,361	447
High License	1881	4,857	337
" "	1882	4,854	190

From the table of local licenses issued in Detroit it appears that there was, from 1882 to 1883, an increase of malt liquor licenses from 208 to 305. Compared with the above figures, what does this prove? Simply that malt liquor licenses, being cheaper by $100 than distilled liquor licenses, are taken out and used as a cover for the sale of distilled spirits. The fact that there is no retrogression in the *production* of malt liquors does not affect our assertion, since it is known that a large quantity of the malt product of the State is shipped across the border.

The decrease in the number of retail dealers of malt liquors is equally great throughout the State, as will be seen from the following:

WHOLE STATE OF MICHIGAN.

Year.	Retail Liquor Dealers @ $25.	Wholesale Liq'r Dealers @ $100.	Brewers less $50.	Brewers more $100.	Retail Malt Liq'r Dealers $20.	Wholesale Malt Liquor Dealers, $50.
1880...	$109,036 53	$4,000 00	$2,416 68	$7,658 33	$8,951 65	$4,629 15
1881...	121,426 05	4,572 92	2,570 84	7,400 00	6,737 53	4,574 99
1882...	121,347 50	5,375 01	2,137 50	8,120 83	3,801 73	4,403 33
1883...	138,221 65	6,616 66	1,654 16	7,591 67	3,467 55	6,022 49

On a smaller scale the law works precisely as the famous English Gin Act did. It provokes illicit selling of ardent spirits, without in any manner affecting the number of drinking-places; it diminishes the consumption of malt liquors; it aggravates the evils of intemperance, and fosters immorality. In large cities, like New York, Philadelphia and Brooklyn, the results would be still more pernicious. High licenses would there practically amount to prohibition, so far as three-fourths of that large class of saloons are concerned in which only malt liquors are sold. They would not affect the two extremes of the business; they would neither diminish the number of gorgeous establishments frequented by the "gilded youth," nor the number of low dens, where profligacy and crime find refuge; but they would undoubtedly decrease the number of beer saloons in those densely populated quarters where cheap and wholesome stimulants form almost the only solace and comfort of the great mass of hard workers and their families. It would almost exclusively affect those citizens who, as has been shown by Mayor Low's inebriety statistics, vie with the best citizens in sobriety and strictest obedience to the law. The system, as a temperance measure, is not needed in those quarters; but where it might be said to be needed, it would have absolutely no effect. Proprietors of low dens will surely not shrink from adding one more wrong to the list of nefarious doings which form the sum of their iniquity. They will either sell illicitly, or pay the high license, and strive to make up the extra expenditure by an extra effort in depravity. They can illicitly sell ardent spirits, because these can readily be concealed or transported from place to place in small quantities. Beer cannot be sold illicitly, as every one knows, without exposing the seller to easy discovery and punishment. In place of the small respectable beer saloons, we would see groggeries in the disguise of soda-water stands, and like seemingly harmless business. Whiskey would be consumed in larger quantities, and under circumstances almost excluding the possibility of preserving public order and morality; while beer would, to a large extent, be driven out in just those quarters where light, wholesome stimulants have become an absolute necessity to the happiness and comfort of a large well-behaved and orderly portion of our population.

This is the lesson that the operation of the Michigan law teaches us. It is needless to cite the examples of other States having high licenses; these two illustrations, one of English, the other of American origin, suffice to convince any fair-minded person, that whenever the license system becomes partial prohibition in disguise, it works quite as disastrously to temperance and morality as prohibition pure and simple.

But, even if a high license system ever could diminish the number of saloons as automatically as it manifestly does the very opposite, it would not necessarily follow that temperance would be the gainer by it. Temperance advocates contend that it would, and in proof of their assertion they cite the Gothenburg system. It is seriously to be doubted whether these persons know the system they praise.

Let us see how the Gothenburg law arose, and what it is.

THE GOTHENBURG SYSTEM.*

Intemperance in Sweden is also a product of defective laws. There was a time, when every Swedish cultivator of the soil had to plant forty poles of hops (1440), and there is sufficient evidence that the brewing industry was subsequently encouraged by the Government. As a nation, the Swedes were not then an intemperate people. Their trouble began, when, in 1787, every family was given the right to distill liquors for their own consumption. In 1800 all restrictions were abolished, and it is stated that the per capita consumption rose thereafter to 29 quarts. In 1829 the number of stills was 173,124, of which 172,043 were in operation in rural districts. Farm-hands were not infrequently paid their wages in spirits, and drinking bouts were thought as much an economic necessity as a matter of appetite and pleasure. Even the Government seemed to take it for granted that a causalty existed between the prosperity of agriculture and stockraising on the one hand, and this universal distillation on the other. The surplus of cereals and fruit for which in years of plenty no market could be found, had to be utilized in some way, and in none, it was thought, more advantageously than by that of transforming it into liquors. The effects of this state of things soon assumed the form of a national calamity, and the Government, now frightened out of its indifference by what Dr. Huss declared to be evidences of a rapid physical and mental decadence, adopted one punitive measure after another in the hope of checking the excesses. Finally the right of distilling was curtailed; the general Government, in 1855, limited the time for distilling to two months in each year, (from the 15th of October

* We make use of the official report relative to "Tilverkning och Försäljning af Bränvin" 1878–79; of a historical sketch on the subject, and the German report: "Die schwedischen und norwegischen Schankgesellschaften." Bremen, 1883.

to the 15th of December;) imposed a uniform tax of 16 shillings on every *kan* of liquor, making no discrimination between liquors distilled for home-consumption and those for the trade; the right to sell from one *kan* up, which had theretofore been vested in every landed proprietor was curtailed by fixing the minimum quantity allowed to be sold at 15 *kans;* only those holding licenses being allowed to sell in less quantities. The granting of licenses was made the prerogative of municipal authorities, under supervision of an officer of the general Government; licenses were to be sold at auction, in such a manner that the person offering to pay tax on the largest quantity of liquor was to obtain the privilege; retailing companies were to be allowed to buy the privilege at a fixed sum. In 1860 the right to distill for home-consumption was entirely abolished and all distilleries were placed under surveillance, and the time for distilling was extended to seven months. In 1867 other laws were passed, the essence of which is stated below. In 1869 a law was passed prescribing the process of rectification, so as to exclude deleterious substances. Under these laws the Gothenburg system grew up, which is to-day in operation throughout Sweden and Norway. The principal features of this system are: I. The number of licenses allowed to be granted is prescribed by law. II. The right to sell under *all* these licenses is sold either at auction, or for a fixed sum to retailing companies, consisting of philanthropists. III. The saloons of the companies are managed by salaried officers, who are forbidden to encourage the guests to drink ardent spirits. IV. In the saloons, beer, tea, coffee, and solid food are furnished at moderate prices; the rooms are airy, well-furnished and provided with reading matter, pictures, flowers, &c.. V. Only perfectly pure and thoroughly rectified spirits are offered for sale. VI. Malt liquors are exempt from the restrictions imposed on the manufacture and sale of distilled liquors. VII. According to a law of 1873, all profits arising from the sale of distilled liquors are paid into the municipal treasury. VIII. The municipal authorities endeavor to ameliorate the condition of the laboring people.*

This is a rough outline of the Swedish system. The law works admirably, but has not until recently reduced the consumption of ardent liquors. Certain classes of the population even drink more of these liquors now than formerly, while other classes are fast becoming accustomed to beer-drinking, as ir evident from the fact

* A committee of the Board of Aldermen of Gothenburg recommended, in 1864, the erection of comfortable dwellings for the workmen.

that the production of malt liquors rose from 52,340,000 litres in 1870, to 97,638,160 litres in 1882. The favorable results of the operation of the law are attributable almost entirely to the better quality of the distilled liquors, to the amelioration of the condition of the working people and a change in the drinking habits of the better classes. We quote from the report of the German Travelling Commission (1883):

The appreciable decrease in the number of cases of public drunkenness and the comparatively rare occurrence of those diseases which are caused by the excessive abuse of alcohol are largely due to the fact, that brandy as it is used now has become better, its quality has improved and it is less injurious to those who drink it.

In former years, the poorer classes and especially the inhabitants of the rural districts—in Sweden as well as in other northern countries used to prepare whiskey from raw spirits and drink it as it came from the still. This liquor, which contained all the different kinds of alcohol, formed during the process of distilling, and which was consumed in great quantities, had that pernicious effect with which Magnus Huss, the celebrated physician of the Seraphine-Hospital at Stockholm, familiarizes us in his excellent work on "Chronic Alcoholism." Later the rectification of brandy as a beverage by means of charcoal was introduced, and the law of March 21, 1869, prohibited the sale and retailing of spirits which were either adulterated by deleterious ingredients, or not well rectified, or contained more than 46 per cent. of alcohol. Even this kind of whiskey, rectified by the so-called cold process, still contains quite a considerable quantity of fusel-oil, and it had long been supposed that these oils, have that destructive effect which assumes the form of chronic alcoholism, and after long continued excesses of delirium tremens. It is true, Huss believed that fusel-oil only served to hasten the injurious effects of pure brandy, but he believed so, because Dahlstrom had not succeeded in producing symptoms of poisoning in animals by introducing fusel-oil into their system. The experiments made by Cros, Rabuteau, Richardson, Dujardin-Beaumetz and Audigé, by Eulenberg, Binz and his disciples, show beyond any doubt, that the poisonous properties of alcoholic drinks depend to a great extent on the quantity and quality of these injurious ingredients. Now, it has been proved that spirits, obtained from grain, turnips or potatoes, contain, besides æthyl-alcohol, considerable quantities of that highly injurious substance called amylic alcohol ; and Isidore Pierre and others have shown that these poisonous substances cannot be removed from the spirits by the charcoal filter, but that a pure drink can only be obtained through repeated distillation by means of the so-called rectifying-process. In potato-spirit, rectified after the "cold system," there were found, according to Rabuteau, at least 5 per cent. of fusel oil ; in brandy, once rectified rapidly, and then purified on charcoal, there were found from 2 to 3 per cent., and of these impure substances, amylic alcohol, a very poisonous stuff, which is at least fifteen times as strong as æthyl-alcohol, formed about two-thirds. The injurious effect of spirits on the human system depends upon the quantity of poisonous substances it contains ; it is for this reason, that alcohol is found to be most deleterious where potato-brandy is being drank, that was either insufficiently rectified, or not at all. In Sweden a rectifying establishment was erected on the improved plan, at Reymersholm, near Stockholm, in 1875, and from that time the

Swedish people succeeded in getting pure spirits. The Commission appointed in 1877, for the revision of the brandy-laws recommended that "liquor offered for sale shall contain no more than 46 per cent. of alcohol; that it must be free from fusel oil, and in every way come up to the standard determined upon by the royal authorities," (following the calorimetric method of Prof. Stenberg at Stockholm, according to which 1 per cent. of amylic alcohol becomes discernible in distilled liquors) "or that brandy must be obtained from raw material thus purified."

In the saloons, managed by the retailing companies, no other but perfectly pure, as they say, "tenfold rectified," liquor is sold, and in the retail-stores spirits purified by means of charcoal, doubly rectified, is sold only when a customer expressly calls for it. Thus, during the week from August 5th to 11th, there were sold in the retail stores at Stockholm 6,005 *kans* of tenfold-rectified liquor and only 14 kans of doubly-rectified splrits, while in the saloons of the retailing company there were sold 9,710 kans of perfectly pure and no doubly-rectified brandy at all. This extraordinary change in the quality of the beverages offered for sale accounts for the favorable results.

The Swedish law undoubtedly deserves to be imitated wherever intemperance prevails to such a degree as to give rise to fears of a degeneracy of race. But, aside from the fact that drunkenness does not now prevail, nor ever has prevailed, to such an extent in our country, it is pretty clear that the Swedish system would prove a total failure in America. With their habitual short-sightedness and superficiality, temperance advocates harp on the one single feature of reducing the number of saloons, ignoring entirely that this reduction has not, until recently, decreased the consumption of spirituous drinks in Sweden, and that all the benefits of the Gothenburg system spring from those features which either could not stand the ordeal of public criticism in our country, or could not be realized on account of a lack of that practical philanthropy which is the main-spring of the Swedish success.

Think of a private company in New York having a monopoly of licenses for the sale of liquors! Think of that monopoly paying its profits on distilled spirits into the public treasury! Imagine, if you can, that this corporation would build airy, comfortable saloons, and furnish its guests with wholesome food and drink at a nominal profit! Then think of our Government maintaining such a system of control and supervision over the manufacture of distilled spirits that adulterations would be excluded! These thoughts seem preposterous in themselves.

Without the change in the quality of the drink; without the efforts of the unselfish public-spirited retailing companies; without the amelioration of the condition of the workers, and without that feature which encourages the use of fermented drinks, the mere reduction of the number of saloons would have accomplished nothing in the interest of temperance.

It would be carrying owls to Athens to recapitulate the showing of our statistical tables, or to elucidate what, in the preceding chapters, has been demonstrated to have been the result of prohibitory and excessively stringent liquor laws. From all that has been said, it follows that, in dealing with this question, law-makers should bear in mind that the use of intoxicants is not a vice, but a perfectly proper enjoyment of great physical, intellectual and moral benefit to the individual, and of inestimable ethical and material advantage to society; that the abuse of inebriating liquors is a vice, and that, while society is warranted in protecting itself against the effects of inebriety, the method of such protection should not in the least affect the liberty of action of the drinker, but should hold the drunkard responsible, not, indeed, for drinking to excess, but for such harm as he may do to others. The wise law-maker should also consider that all laws aimed at the proper use of intoxicants are unjust, because detrimental to the happiness and well-being of the majority; that they are immoral, because they create a host of law-breakers, corrupt morals, aggravate the evils of intemperance by driving the people to the use of ardent spirits, and by placing the drinker in circumstances which inevitably lead to excesses; that they are tyrannical, because they infringe upon the personal liberty of all; that they are economically pernicious, because they aim to destroy many branches of industry and agriculture, in which enormous sums of money are invested, and upon which thousands of skilled and unskilled laborers depend for a livelihood; and, finally, that they are socially dangerous, inasmuch as they do away with many opportunities for recreation and popular amusement, and thus destroy the very means by which social advancement and refinement can be effected.

The discrimination between use and abuse, drinker and drunkard, should determine the character and scope of the laws for the regulation of the liquor traffic; for if it be perfectly honorable and legitimate to drink intoxicants, it must of necessity be honorable and legitimate to sell them; hence, it is unjust to the last degree to place the upright, law-abiding citizens, who sell distilled or fermented liquors, on the same legal footing with the proprietors of disreputable drinking places. The evil influence of disreputable resorts upon the safety and morality of society cannot be paralyzed by declaring the whole traffic to be dishonorable, nor will it meet the ends of justice and the requirements of society to persecute the innocent on the pretense of restraining the guilty. Experience teaches us that such a course invariably injures the honorable dealer, while it positively benefits the guilty, and consequently aggravates the evils against which it is alleged to be directed. The fiscal policy in reference to the sale of intoxicating drinks should be so regulated as to promote temperate drinking habits and diminish drunkenness, and this, as we have shown by scores of examples, can be accomplished by favoring all the milder intoxicants, without, however, placing unreasonable restrictions upon the sale of ardent liquors.

Under the operation of such laws temperance will flourish without other external aid than that, which may be expected from the influence of good examples and the power of moral suasion.

APPENDIX A.

CORRESPONDENCE

RELATIVE TO INSANITY

CAUSED BY

INTEMPERANCE.

APPENDIX A.

CORRESPONDENCE

RELATIVE TO INSANITY

CAUSED BY

INTEMPERANCE.

I

ALABAMA INSANE HOSPITAL,

TUSKALOOSA, ALA., January 29, 1884.

DEAR SIR: Yours of the 22d inst. received. Upon examination of our books I find there were admitted to this hospital during the last twelve months 210 patients —150 men and 60 women. Of the men insanity was alleged to have been caused by the excessive use of alcoholic liquors in nineteen (19) cases. Among the women there was no case of alcoholic insanity. The nineteen cases among the men was the direct result of the abuse of alcohol, or rather so reputed; how many of the others were *indirectly* due to the same cause it would be difficult even to conjecture.

I am very respectfully yours,

T. BRYCE,
Superintendent.

II.

NAPA STATE ASYLUM,

NAPA, CALIFORNIA.

Year.	Number of Patients Admitted.	Number of Cases of Insanity caused by Intemperance.	
		MALE.	FEMALE.
1879	217	25	4
1880	572	31	7
1881	563	31	5

III

INSANE ASYLUM OF THE STATE OF CALIFORNIA.

STOCKTON, CAL., December 16, 1883.

DEAR SIR: In consequence of the migratory habits of the people of this State the records of this Asylum are necessarily very incomplete as to the causes of insanity, and the percentage which may be attributed to alcoholic drinks cannot accurately be determined. The report of Dr. G. A. Shurtleff, which I send by to-day's mail, contains all the information I can give you on the subject.

Respectfully yours,

W. T. BROWNE.

STOCKTON STATE ASYLUM,

STOCKTON, CALIFORNIA.

Year.	Number of Patients Admitted.	Number of Cases of Insanity caused by Intemperance.	
		MALE.	FEMALE.
1878	219	19	5
1879	106	8	..

IV.

COLORADO STATE LUNATIC ASYLUM,

PUEBLO, COL., December 10, 1883.

DEAR SIR: Having just moved into a new hospital building and being without clerk or assistant, I have been unable to reply to your favor in the manner desired. I may state, however, from the register book of patients received where the existing cause is entered, if known, that it appears the excessive use of alcoholic beverages is assigned as the cause *with ten per cent.* of the admissions here during the past two years. Hoping you will pardon my delay in replying, "as well as brevity," for reasons stated above.

I remain very respectfully yours,

T. R. THOMBS.

V.

CONNECTICUT HOSPITAL FOR INSANE,

MIDDLETOWN, November 23, 1883.

DEAR SIR: Statistics respecting alcoholism are very unreliable. Friends of patients rarely tell the truth. They rarely—almost never—assign alcoholism as the cause of insanity in the cases which we find are caused *solely* and *purely* by intemperance. Experience shows us, that *at least* twenty per cent. of all cases are caused by alcoholism.

Yours truly,

A. M. SHEW.

VI.

RETREAT FOR INSANE,

Hartford, Conn.

Year.	Number of Patients Admitted.	Number of Cases of Insanity caused by Intemperance.	
		MALE.	FEMALE.
1850	135	1	..
1851	128	7	2
1852	158	5	..
1853	140	3	..
1854	177	4	2
1855	169	4	..
1856	157	8	..
1857	161	2	2
1858	144	4	..
1859	141	6	1
1860	168	15	3
1861	164	5	2
1862	171	4	..
1863	170	10	2
1864	143	13	1
1865	155	10	1
1866	165	6	..
1867	182	4	3
1868	173	7	6
1869	120	3	1
1870	123	10	2
1871	143	8	1
1872	115	8	3
1873	114	7	1
1874	83	5	..
1875	78	6	..
1876	103	7	1
1877	92	7	2
1878	89	2	2
1879	78	3	2
1880	108	7	1
1881	114	4	2
1882	64	5	..
1883	78	4	1

To 1869 there was no other institution for the insane in the State. Since that time it has received patients from every portion of the country.

H. P. STEARNS, M. D.

VII.

LITCHFIELD, CONN., December 3, 1883.

DEAR SIR: My establishment is so small (being a strictly private one) and the number of cases of alcoholism but a fractional number of all that I receive, that no general conclusions can be drawn therefrom. It seems to me that such inquiries as you propound are applicable to large public institutions.

Respectfully yours,

H. W. BUEL, M. D

VIII.

GOVERNMENT HOSPITAL FOR THE INSANE,

WASHINGTON, D. C.

Year.	Number of Patients Admitted.	Number of Cases of Insanity caused by Intemperance. MALE.	FEMALE.
1855	63	1	..
1856	47	3	..
1857	52	1	..
1858	43	2	..
1859	65	9	..
1860	92	6	1
1861	95	3	..
1862	186	5	..
1863	355	..	..
1864	509	11	..
1865	514	5	1
1866	222	10	..
1867	109	4	..
1868	153	3	2
1869	168	4	..
1870	182	3	..
1871	195	2	..
1872	186	4	..
1873	204	3	1
1874	230	11	..
1875	230	4	..
1876	213	20	..
1877	198	43	3
1878	182	23	3
1879	222	28	1
1880	225	54	3
1881	223	32	1
1882	247	36	..
1883	265	54	1
Total	5675	384	7

GOVERNMENT HOSPITAL FOR THE INSANE,

WASHINGTON, OCT. 29, 1883.

DEAR SIR: I have to acknowledge the receipt of your communication of the 25th inst., and in response thereto enclose herewith the transmitted blank filled out from the records of this Hospital. A few explanations may be necessary:—This hospital was instituted for the care and treatment of the insane from the Army, Navy, Marine Corps, Revenue Cutter Service and the *indigent* insane of the District of Columbia. Pay-patients from the latter source are received only when our accommodations permit; for all practical purposes it may be assumed, that all the insane from the District are provided for here. The first patient was received January 15, 1855, and the institution being directly under government control, reports are made to the end of each fiscal year, to wit: June 30th. *All cases in which the assigned cause was intemperance in the use of alcoholic stimulants have been included, no matter what form the mental disease may have assumed.* Of the 5,675 cases treated, 4,610 have been males, 1,065 females, of which number there remained under treatment June 30th, 1883, 994; males, 755; females, 239.

Very respectfully,

W. W. GODDING,

Superintendent.

IX.

STATE LUNATIC ASYLUM OF GEORGIA,

NEAR MILLEDGEVILLE, GA., NOV. 30, 1883.

DEAR SIR:—In reply to your letter making enquiry as to the per cent. of mental disturbances in our Institution, produced by the use of "Alcoholic Beverages," I must say that it is impossible to give a correct statement, as the parties accompanying patients to the Institution frequently know nothing of their history, or of their antecedents. Hence, you can readily see, it is impossible to keep a correct record of the causes. I also frequently find relatives are loth to *speak of hereditary tendencies.* While I think, perhaps ten per cent. would cover all the cases in our Institution, the *exciting cause* being alcoholic stimulants, I regard alcoholic stimulants as one of the most potent predisposing causes of mental maladies. The descendants of drunkards, or drinking parents are much more liable or susceptible to insanity, whether they themselves are in the habit of drinking or not, than are the children of sober parentage. Imbecility, epilepsy, insanity, etc., etc., may be traced to the intoxicated brain of a parent, or even more remotely their grandparents. If I am correct in my views, it would be difficult to obtain only the direct or exciting cause, which is a small per cent. compared to the predisposing cause. I send you one of our last reports.*

Respectfully,

T. O. POWELL,

Superintendent.

*The report referred to contains no information relative to the cause in question.

X.

ILLINOIS CENTRAL HOSPITAL FOR THE INSANE,

JACKSONVILLE, ILL.

Year.	Number of Patients Admitted.	Number of Cases of Insanity caused by Intemperance.	
		MALE.	FEMALE.
1851	34	1	1
1852	110	4	..
1853	145	6	..
1854	122	4	..
1855	132	2	..
1856	156	3	..
1857	170	6	1
1858	142	..	1
1859	180	8	..
1860	157	6	..
1861	178	5	1
1862	207	4	1
1863	183	2	..
1864	224	3	..
1865	206	3	..
1866	248	1	..
1867	293	6	..
1868	328	2	1
1869	353	11	..
1870	350	12	..
1871	343	10	..
1872	278	7	1
1873	283	10	..
1874	197	6	..
1875	310	7	..
1876	248	10	..
1877	329	14	..
1878	281	7	1
1879	237	7	1
1880	251	6	..
1881	244	6	..
1882	265	11	..
1883	215	8	..

XI.

ILLINOIS SOUTHERN HOSPITAL FOR THE INSANE,

ANNA, ILL.

Year.	Number of Patients Admitted.	Number of Cases of Insanity caused by Intemperance.	
		MALE.	FEMALE.
1874	158	2	..
1875	103	4	..
1876	147	2	..
1877	92	1	..
1878	308	10	.
1879	190	3	1
1880	138	4	1
1881	130	1	..
1882	174	2	..
1883	157	1	..
TOTAL,	1,597	30	2

XII.

ILLINOIS EASTERN HOSPITAL FOR THE INSANE,

KANKAKEE, ILLINOIS.

Year.	Number of Patients Admitted.	Number of Cases of Insanity caused by Intemperance.	
		MALE.	FEMALE.
1879	22	7	.
1880	139	25	..
1881	278	24	..
1882	139	14	1
1883	188	17	2

KANKAKEE, NOV. 13, 1883.

DEAR SIR :—The statistics includes information in about 50 % of all our cases; the remainder we have no history of. The apparent larger proportion during '79, '80, '81, is due to our having gotten more information than we are able to obtain on recent admission. The number *in toto* seems small, but you may be aware that we do not admit *inebriates*. The figures do not include re-admissions of which there are a number each year. The name of each is counted but once. The left hand column gives the number admitted each year, not counting those who have been intemperate and already residents of the Hospital.

Very truly,

R. S. DEWEY,

Superintendent.

XIII.

COOK COUNTY HOSPITAL FOR THE INSANE,

JEFFERSON, ILLS., DEC. 31, 1883.

DEAR SIR:—I handed your letter to our assistant and expected it had been attended to, but find it has not. I do not think that alcoholic stimulants are the *direct* cause of insanity, except in a small percentage of cases, and that when the taste has become morbid and is indulged in to an unreasonable extent. They are probably the *exciting causes* in every fourth case in this asylum. I mean that with a tendency to insanity the injudicious use may prove an exciting cause and frequently does. But it is not the stimulants, except in connection with the predisposition to the nervous trouble. We have about three fourths of a million population in this county and receive our patients from this county alone. The number of inmates in this asylum in 1883 was 447, viz.: 232 males and 215 females.

Very respectfully,

J. G. SPRAY,

Superintendent.

XIV.

INDIANA HOSPITAL FOR INSANE,

INDIANAPOLIS, IND.

Year.	Number of Patients Admitted.	Number of Cases of Insanity caused by Intemperance.	
		MALE.	FEMALE.
1871	338	14	3
1872	312	8	1
1873	320	12	1
1874	373	16	2
1875	438	17	..
1876	489	14	..
1877	477	16	..
1878	470	15	1
1879	615	12	..
1880	914	9	1
1881	728	10	2
1882	762	22	4
1883	698	13	2

This Hospital was opened November 1, 1848. From that date to October 31, 1870, there were 4,431 persons admitted. In that number there were 147 cases attributed to alcoholism,—143 men and 4 women. From that last date I give you only the admissions for each year as that is the correct basis of calculation. Also, number of cases of alcoholism each year of men and women.

WM. B. FLETCHER,

Superintendent.

INDIANOPOLIS, IND., NOV. 2, 1883.

MY DEAR SIR:—By direction of Dr. W. B. Fletcher, Superintendent of this Hospital, I have filled the enclosed blank-form and return it to you. When any one studies the subject of insanity by practical observation in a Hospital for the Insane, it will be astonishing how little of this disease is produced by the use of alcoholic drinks. But it is assumed by certain persons that fifty per cent. of nervous diseases and insanity is the direct result of an indulgence in such beverages. This is all wrong.*

Very truly,

A. J. THOMAS,

First Assis't Physician.

* The compiler invites attention to an article by Dr. Thomas, reprodnced in the text of this sketch.

XV.

IOWA HOSPITAL FOR THE INSANE,

MOUNT PLEASANT.

DEAR SIR:—Your communication of the 15th inst., is before me.

I will say that in the eighteen years that I have been connected with Hospitals for the Insane, by observation and preparing statistics, I am prepared to say that about twenty five per cent. of the causes of insanity, in something more than seven thousand patients that I have been familiar with, has resulted from the use of alcoholic beverages. This refers only to those where that is the *direct* cause. From the history of other cases, I judge that indirectly twenty-five per cent. more may be traceable to the same cause, as a result of drunkenness in the parents, affecting the nervous organization of the children; and in the case of many females, the hardships and cruelties which they undergo is a result of this excess.

Very truly yours,

H. A. GILMAN,

Superintendent.

XVI.

IOWA HOSPITAL FOR THE INSANE

AT INDEPENDENCE.

Year.	Number of Patients Admitted.	Number of cases of Insanity caused by Intemperance.	
		MALE.	FEMALE.
1874	239	6	..
1875	411	9	..
1876	472	10	1
1877	541	4	1
1878	577	5	..
1879	732	9	1
1880	705	10	..
1881	819	10	1
1882	820	14	..
1883	871	18	1

XVII.

KANSAS STATE INSANE ASYLUM,

Superintendent's Office,

TOPEKA, KAN., November 22, 1883.

DEAR SIR: Your communication of the 15th inst. is before me. We did not fill out or send you the statistics you called for, because we could not do so in a way to make them of any use. We spent some time in working up the matter; our difficulty was in fixing upon the population represented in this Institution at the time. There are two asylums in the State, and no special territory is designated for either. Then, a part of the time but few new cases were admitted, and those carried over from one year to another would, of course, be counted a second time.

Again in many cases there is a combination of causes, so that when one thing is assigned as a cause, something else could as well be assigned in the same case.

We receive cases every year where the excessive use of alcoholic drinks has very decidedly been the primary cause, as well as many where it is only the exciting cause. As to the figures, the percentage of cases of this character much depends upon the point to be made and how carefully the cases are analyzed.

This is not very satisfactory to you, I know, and is not to me, but you say "a reply, no matter of what kind."

Yours truly,

A. P. TENNEY.

XVIII.

KANSAS STATE INSANE ASYLUM,

OSAWATOMIE, December 11, 1883.

DEAR SIR: Yours of the 30th ult. received. Your previous letters of October 22nd and November 16th came in due time. I must ask your pardon for delay of answer. Since the 15th ult. I have been deprived of the services of the Assistant Superintendent until the 7th inst., and, as a consequence, so many extra duties have devolved upon me, and my whole time was so constantly occupied, that I have been unable to devote any attention whatever to the subject matter of your letter. In regard to the data you desire, I am sorry to say I am unable to give you anything reliable or satisfactory from our records, some of which were destroyed by fire three years ago. Our history of cases is made up from the best information we can obtain, just previous to or at the time of admission. In many cases little of their history can be ascertained. To the question, "Was the patient ever addicted to intemperance in any form?" we often receive answers which are disproved or rendered doubtful by subsequent information. In the absence of reliable data on the subject, I can hardly believe a conjecture would be of any value. A careful study of the direct and indirect influence of alcoholism in the causation of mental derangement is required to arrive at even an approximation of the true relation of cause and effect. This opens up a wide range of enquiry upon a very important question, and I am free to confess that I am not prepared with the requisite data to defend and sustain an opinion, were I to venture one, as the result of personal observation.

Very truly yours,

A. H. KNAPP,

Superintendent.

XIX.

33 Market St., Lexington, Ky., Dec. 26, 1883.

Dear Sir :—I am no longer Superintendent of the Eastern Kentucky Lunatic Asylum, and have not been for several months, consequently cannot give you the statistical information you desire as the records are no longer under my control. I have sent your letter or letters to the present incumbent, Dr. R. C. Chenault, and would advise that you open a correspondence with him. Accept my apologies with the above explanation for my apparent neglect, and believe me,*

Very respectfully yours,

W. O. BULLOCK.

* No reply received from Dr. Chenault.

XX.

WESTERN KENTUCKY LUNATIC ASYLUM,

Hopkinsville, Ky., January 14, 1884.

Dear Sir :—An approximate estimate of *intemperance* as a cause, direct or indirect, of insanity in this institution is seven and one half per cent. Asylums have only been assigned districts in past few years, hence can not give you population from year to year. This Asylum was opened in 1854, has treated more than two thousand four hundred patients, or in full numbers two thousand four hundred and sixty-eight up to date.

Respectfully yours,

JAS. RODMAN,
Superintendent.

XXI.

MOUNT HOPE RETREAT,

Maryland.

Year.	Number of Patients Admitted.	Number of cases of Insanity caused by Intemperance.	
		MALE.	FEMALE.
1883	585	36	18

In reference to the subject of intemperance we find on consulting the statistics that fifty-four of the persons treated were acknowledged to use liquors and tobacco in excess, and it was to the excessive use of these stimulants that the friends ascribed the foundations of the nervous difficulties which resulted in insanity. Thus abundant proof is at hand to show that intemperance does have a direct influence upon the production of this disease. Indeed every year's experience tends to demonstrate most conclusively, that intemperance itself does produce insanity, even when no predisposition to the disorder exists ; but, when we add to this habit the condition known as an insane temperament, or direct hereditary predisposition, the individual is doubly liable to an outbreak of mental disorder, which experience demonstrates is more persistent, and less liable to end in recovery than most of the so-called curable forms of insanity. When we think of the dire calamities and untold miseries likely to result from an attack of insanity produced by these causes, what a powerful incentive should it be to do everything possible to escape an attack.*

* This is an extract from the physician's annual report.

XXII.

BAY VIEW ASYLUM,

DEPARTMENT CITY ALMSHOUSE,

BALTIMORE Co , December 27, 1883.

In answer to both your favors I would say that the majority of our insane are received with no history and consequently our records do not enable me to give the information you desire.

Very respectfully, JOS. T. BARTLETT,
Res. Physician.

XXIII.

ASYLUM FOR THE CHRONIC INSANE,

WORCESTER, MASS., November 19, 1883.

I am unable to answer your questions for the reason that all the patients at this Asylum are transfers from other hospitals in the State, and we therefore know little or nothing of their previous history.

Yours respectfully,
H. M. JUMBY.

XXIV.

McLEAN ASYLUM,

SOMERVILLE, MASS., November 17, 1883.

DEAR SIR:—An examination of our books shows that a statement of numbers as called for by your blank form would lead to such erroneous conclusions that the information would be of no value.

During the first ten or fifteen years of the period since 1850, numbers of persons were admitted here of their own wish, with alcoholism ascribed as the cause of their temporary mental disturbance, which was not strictly insanity. Several persons came here for that cause a number of times in each for a number of years. It is impossible to give correct dates eliminating these errors that would thus be caused in a statement of numbers.

This practice was finally discontinued, and of late years, I cannot say how many, only cases have been admitted with insanity as caused by alcoholism in which insanity has really appeared and continued for a length of time.

I think that also of late years alcoholism may not have so readily been accepted as a possible cause of insanity in our cases.

For these reasons there would be a large apparent reduction in the number of cases of "alcoholism" without their having been any real reduction; this would form a source of error which cannot now be eliminated. For example: in 1850 the total number of cases treated in the Asylum during the year was 357; of these, 30 are recorded as cases of alcoholism. In 1882 the corresponding figures are 237 and 9. While I have no doubt insanity is caused by excessive use of alcoholic beverages I am not prepared to express a decided opinion as to the greater or less prevalence of insanity due to that cause. In the class of cases received here we do not see much of it.

Very truly yours, EDWARD COWLES,
Superintendent.

XXV

BOSTON LUNATIC HOSPITAL.

Year.	Number of Patients Admitted.	Number of Cases of Insanity caused by Intemperance.	
		MALE.	FEMALE.
1881	54	2	..
1882	115	9	1
1883	104	10	1

Intemperance assigned as a cause on certificate, including cases of alcoholism.

THEO. FISHER,
Superintendent.

XXVI.

TAUNTON LUNATIC ASYLUM,

Taunton, Mass.

Year.	Nomber of Patients Admitted.	Number of Cases of Insanity caused by Intemperance.	
		MALE.	FEMALE.
1854	330	24	4
1855	406	10	4
1856	447	15	7
1857	505	37	12
1858	550	31	36
1859	532	40	9
1860	586	43	11
1861	613	52	9
1862	619	37	12
1863	617	35	16
1864	605	31	12
1865	560	14	10
1866	551	30	9
1867	606	29	12
1868	649	35	7
1869	663	35	5
1870	758	29	9
1871	762	20	3
1872	828	30	8
1873	845	26	6
1874	889	29	20
1875	985	35	14
1876	1185	50	16
1877	1244	28	18
1878	1208	21	8
1879	752	14	4
1880	749	16	3
1881	828	26	5
1882	786	25	5

TAUNTON, MASS., November 20, 1883.

DEAR SIR: I enclose you the table of statistics which you ask for, after some unavoidable delay which I regret. It is only approximately correct, but perhaps is as nearly so as such tables usually are. It shows the number of insane persons admitted to the hospital in whom intemperance was the exciting, or one of the exciting causes of their insanity. It includes cases of alcoholism properly so called, and also those whose mental disturbance of whatever type was supposed to have resulted from the excessive use of alcoholic liquors.

Yours truly,

W. BROWN

XXVII.

DANVERS LUNATIC HOSPITAL AT DANVERS, MASS.

Year.	Number of Patients Admitted.	Number of Cases of Insanity caused by Intemperance.	
		MALE.	FEMALE.
1878	222	25	11
1879	533	44	22
1880	607	13	13
1881	626	59	20
1882	656	73	23
1883	721	64	24

XXVIII.

EASTERN MICHIGAN ASYLUM,

PONTIAC, MICH., November 1, 1883.

DEAR SIR: Enclosed please find the blank which you sent duly filled out. This Asylum was opened in 1878, and consequently no figures can be given back of that period. During the first year a large number of patients were received immediately from the Michigan Asylum for the Insane at Kalamazoo and a relatively large number of alcoholic cases. By the term "Alcoholism," I understand what I have always denominated Alcoholic Insanity. Under this term are included three distinct forms, which I will briefly describe as follows:

1. *Mania a potu*, or maniacal excitement caused by excessive alcoholic indulgence. It is of brief duration, and usually terminates in recovery. Where it does not, a condition is developed which constitutes a second variety.

2. *Chronic Alcoholism.* This is characterized by hallucinations of sight and hearing, apprehensions of persecution and homicidal impulses. There are marked irritability, a malignant ingenuity in fault finding, and a complete lack of appreciation of condition.

3. The third variety is *dipsomania* proper. In this there is well marked depression, followed by an intense craving for alcoholic indulgence, lasting, usually, for a comparatively brief period, followed by abstinence and correct life, until the appetite for the vicious indulgence is re-awakened by a fresh attack of depression.

If you desire to have these cases discriminated, and to include merely cases of chronic alcoholism, such as are described under the second head, I can, of course, make the distinction. It seems to me, however, that they can be more correctly included under the general term alcoholic insanity, since the mental disease in these cases is directly to be traced to alcoholic indulgence.

Very truly yours,

HENRY M. HARD,
Medical Superintendent.

Year.	Number of Patients Admitted.	Number of Cases of Insanity caused by Intemperance.	
		MALE.	FEMALE.
1878	313	20	4
1879	490	10	1
1880	567	8	2
1881	600	11	2
1882	658	12	1
1883	806	21	6

XXIX.

MICHIGAN STATE RETREAT,

DETROIT, MICH.

Year.	Number of Patients Admitted.	Number of Cases of Insanity caused by Intemperance.	
		MALE.	FEMALE.
1883	238	102	12

The institution is self-supporting and is owned and conducted by the Sisters of Charity.*

* The figures are not included in Table I, because the institution may, in a certain sense, be classed with inebriates' homes.

XXX.

MICHIGAN ASYLUM FOR THE INSANE.

KALAMAZOO, November 23, 1883.

DEAR SIR: Yours of the 15th inst is received. In reply would say, that our observations do not go to sustain the views entertained by many, that the excessive use of alcoholic beverages is a frequent exciting cause of mental disturbance; in other words, mental disease. A glance at the table of causation, as printed in our last report, shows that out of 3534 patients, the exciting cause of only 159 was attributed to the excessive use of narcotics and alcoholic drinks. While the table is not absolutely correct, still we think it approximates the truth.

The evil, however, resulting from excessive use of alcoholic beverages, cannot be estimated by the number who are made insane, but in order to fully appreciate the extent of mischief done, we must study its remote effects upon the race. I cannot, perhaps, better illustrate my meaning than by quoting from one of our reports. Several years since a young man was admitted into this Institution who had committed homicide while laboring under an attack of maniacal excitement; he had always been a quarrelsome boy and had led a very irregular life. The following facts were ascertained in reference to the family history: His grandfather had been an influential citizen in the city where he lived, and had held positions of honor and trust. He was a man of sound health both mental and physical, but a high liver, and irregular and intemperate in his habits. His children possessed less vigorous constitutions and were in every respect his inferiors. The father of the patient was intemperate, quarrelsome, and shiftless. Several of the grandchildren have been insane, and two inmates of this Institution.

In this case the impairment of the nervous energy by excesses of various kinds, while producing marked effects upon the vigorous constitution of the grandfather, was made apparent in the children by general mental and physical enfeeblement, as well as moral degeneracy, and finally reached its full development in the grand-children, in whom it assumed the form of insanity.

Many of the epileptics, idiots, and insane in the land, are offsprings of drunken parents, to say nothing of those defectious beings found in our penitentiaries.

In conclusion permit me to say that in my opinion alcohol is the greatest enemy of mankind.

Yours very truly,

GEO. C. PALMER,
Medical Superintendent.

XXXI.

MINNESOTA HOSPITAL FOR INSANE,

ST. PETER, MINNESOTA.

Year.	Number of Patients Admitted.	Number of Cases of Insanity caused by Intemperance. MALE.	FEMALE.
1867	97	2	1
1868	47	..	1
1869	130	5	1
1870	143	5	..
1871	139	16	1
1872	118	9	..
1873	140	13	2
1874	194	20	..
1875	188	19	1
1876	253	27	1
1877	237	17	2
1878	249	21	..
1879	248	21	2
1880	232	20	..
1881	172	16	..
1882	267	28	..
1883	253	26	..

C. W. BARTLETT, M. D.,
Superintendent.

XXXII.

MISSISSIPPI STATE LUNATIC ASYLUM.

JACKSON, MISS., October 29, 1883.

DEAR SIR: In reply to your circular letter I will answer to the best of my ability. It is quite difficult to ascertain the causes of insanity in this State. About one-fourth are reported as being unknown.

Very respectfully yours,

T. J. MITCHELL, M. D.
Superintendent.

Year.	Number of Patients Admitted.	Number of Cases of Insanity Caused by Intemperance.	
		MALE.	FEMALE.
1855	51	3	..
1856	64	0	..
1857	71	1	1
1858	90	1	1
1859	109	2	..
1860	115	..	..
1861	125	2	..
1862	131	..	..
1863	130	..	..
1864	106	..	..
1865	112	..	..
1866	119	1	..
1867	146	..	..
1868	135	..	..
1869	148	..	..
1870	155	3	..
1871	162	..	..
1872	227	3	..
1873	304	2	..
1874	305	..	..
1875	321	..	..
1876	325	1	..
1877	340	4	..
1878	392	2	..
1879	394	3	..
1880	402	2	..
1881	439	5	..
1882	420	2	..
1883	429	1	..

XXXIII.

MISSOURI STATE LUNATIC ASYLUM No. 2.

This report is given from our Biennial reports as reported to the General Assembly of the State.

If your object is to show what proportion of insanity alcoholic drinks, used to excess, produces, Asylum reports cannot be relied upon, because often insanity is the cause of excessive drinking, and Superintendents have no means of ascertaining the facts. Another unsettled and important point in relation to this vital question should be considered: how many weary and exhausted brains are saved from disease and ruin by the proper use of alcoholic stimulants?

Very truly,

GEO. O. CORTLETT, M. D.,
Superintendent.

Year.	Number of Patients Admitted.	Number of Cases of Insanity Caused by Intemperance. MALE.	FEMALE.
1876	293	7	..
1878	286	13	2
1880	138	6	..
1880	316	9	2

XXXIV.

STATE LUNATIC ASYLUM,

FULTON, MO., December 5th, 1883.

Yours of the 30th ult., received. Your previous letters were also received, but I was quite unwell at the time and hence failed to respond. Although I have been connected with this Institution nearly 26 years I regret to state that I have not at present a complete file of the reports from the opening of the Asylum in 1851. I was here from the opening of the Institution till 1865, when I resigned on account of impaired health and returned again in January, 1873. During the period of my absence the reports were in some way so scattered that I was able to obtain only one nearly complete file, and this is now in the hands of a gentlemen who is preparing a history of this State. I do not know his present locality. The number of cases of insanity in this Institution that have resulted from the excessive use of alcoholic beverages has been small compared with a majority of other like institutions. The very large majority of our patients are from the rural districts where intemperance does not to any great extent prevail. There are the hospitals for the insane in St. Louis, one St. Louis County Asylum, designed to accommodate all the insane poor in city and county, and one under the control of the Catholics for private or pay patients. We have only two State Institutions, the one here and one at St. Joseph, Mo. Most of our hospitals for the insane, largely filled from the cities, report a large per cent. from alcoholism. Since my return, January 1873, we have admitted (2,000) two thousand and ten patients, and from the best history we could obtain at the date of their admission only seventy seven of the whole number resulted from intemperance. The whole number admitted since the opening of the Asylum has been 3,976. The Institution when first opened only accommodated 80 patients, and we now have 521, but are overcrowded. The building had reached its present capacity before I resigned in 1865. The per cent. from intemperance of the 1,966, after substracting the 2,010, was perhaps a little larger than that, as we admitted patients from St. Louis part of the time. If I had any file of reports, I could give you an accurate statement.

Very respectfully, J. R. P. SMITH,
Supt. Physician.

XXXV.

ALEXIAN BROTHERS' HOSPITAL AND INSANE ASYLUM,

St. Louis, December 8, 1883.

Dear Sir:—Your letter enclosing blank came to hand in due time. It is only a couple of years since we opened our Miserecordia Asylum and have kept no special record of it. Our main building being the *general hospital* for ailments of *all descriptions*, it is not possible, except with a great amount of labor, to comply with your request. During the present year we have erected an addition which enables us to accommodate forty patients, whereas heretofore we had room for only twenty in the Insane Department. At present we have 18 patients, only one of whom can be said to owe his affliction to the excessive indulgence in intoxicating drinks. During the entire year we had ten patients who became such through excessive drinking.

Respectfully yours,

BROTHER JOCUNDUS SCHIFFER.

XXXVI.

ST. LOUIS INSANE ASYLUM,

St. Louis, Mo.

Year.	Total number of Admissions.	Number of cases of Insanity caused by Intemperance.	
		MALE.	FEMALE.
1883.	2,204	169	41

From the opening of the Asylum, April 23rd, 1869.

XXXVII.

NEBRASKA ASYLUM FOR THE INSANE,

Lincoln, Neb.

I cannot give the statistics asked for. I know it is bad enough, bad beyond comprehension.

H. MATHEWSON.

XXXVIII.

NEW HAMPSHIRE ASYLUM FOR THE INSANE,

Concord, October 29, 1883.

Dear Sir:—I regret very much that our records are not sufficiently explicit on the point in question to be of any especial service to you, otherwise I should have been most glad to send them to you.

Very truly yours,

C. H. BANCROFT.

XXXIX.

NEW JERSEY STATE LUNATIC ASYLUM,

TRENTON, N. J., January 4, 1884.

DEAR SIR:—In reply to your favór, I would state that we have at present under care iu our institution six hundred and thirty-three patients—three hundred and thirteen men and three hundred and ninety women. Of the whole number we have only eleven, possibly twelve, that are strictly cases of alcoholism. Of persons whose parents (one or both of them) are intemperate, we have a large number under care.

Very truly yours,

J. W. WARD.

XL.

PASSAIC COUNTY LUNATIC ASYLUM.

Year.	Number of Patients Admitted.	Number of Cases of Insanity caused by Intemperance.	
		MALE.	FEMALE.
1883.	43	10	29

WM. NELSON,
Clerk.

XLI.

BURLINGTON, N. J., December 3, 1883.

DEAR SIR:—I am much interested in your investigation, and await the result with considerable concern.

What the people want is truth, and I am glad to find that in your letter you do not indicate anything like a bias iu any particular direction, but leave the matter entirely open to the construction of your correspondents. It would be more in accordance with science, and really more likely to eliminate the exact truth if your inquiries were conducted upon a specific basis.

Many persons who have charge of the insane, especially in almshouses, are likely to judge hastily and incorrectly. If they find that an insane person has ever used intoxicants, they may decide without further investigation that the insanity has been caused by the drink, while the real truth may be, as I am confident it is iu many cases, that both insanity and inebriety belong to the same class of disorders, and that they both spring from a family or constitutional tendency to insanity or to drunkenness. Insanity is often the cause of inebriety, and so inebriety is the cause of insanity, especially in cases in which there is a family tendency to either or both, and I believe that so inflexible is this law of heredity that in considering the causation of insanity, the limitation of inebriety as a cause may be traced to those in whom exists this constitutional tendeucy. As there is no hope of securing answers to your inquiry, based upon this fact, I fear you will not get actual results that will be satisfactory to scientific men, although the answers to your general question may be such as shall encourage further enquiries based upon a more exact and scientific standard.

Yours, etc.,

JOSEPH PARRISH, M. D.

PEMBERTON, Dec. 3, 1883.

DEAR SIR: The letter and blank of the Brewers' Association enclosed with yours just received. In reply there is not a record in Burlington Asylum to throw any light upon the question at issue.

I do not recall a single case where insanity is caused by alcohol—I know nothing of their family history. Much might be learned of the more intelligent patients by oral questions.

If you will appoint a time I would be pleased to accompany you through the institution, and see what might be gathered in reference to it, from the patients themselves.*

Very truly yours,

To DR. JOSEPH PARRISH. E. HOLLINGSHEAD.

* No data could be obtained from this Institution.

XLII.

NEW JERSEY STATE ASYLUM FOR THE INSANE,

MORRIS PLAINS, N. J., December 26, 1883.

DEAR SIR: I have received your several favors asking for the number of cases of "alcoholism" that have been received and treated in this asylum of late, as compared with those of an earlier period. In answering your enquiry, I desire first to express my regret that so much delay has occurred in my response to your very courteous letter.

This has arisen entirely from the impression, after examining records of cases, that it would be impossible to arrange statistics on the subject sufficiently accurate to serve any useful purpose. The grounds of inaccuracy are as follows:

First, the impossibility of obtaining information as to the *cause* or causes of cases of mental derangement in *many* instances. Second, its *unreliable* character when received from very ignorant or untruthful friends. Third, the uncertainty of the character of the disease, whether from alcoholic stimulants alone, or from the effects of several causes combined to develop it, in which case it might properly be called an ordinary case of insanity. With one set of observers it might be regarded in this way, and by another in the opposite. Again the policy and practice of institutions, corporate, state and city, vary so much in regard to the admission of this class as insane, that it is impossible to compare the cases of one with the other.

While city institutions are compelled to receive all who are lawfully committed to them, state and corporate, may and do exercise a degree of discretion in some cases. Again, while some of the latter institutions choose to receive all of this class of persons for whom application is made, others desire to exclude them to the greatest possible extent. In view, then, of these and other reasons that might be stated, and believing that any statistics of the kind you ask would be inaccurate, and therefore misleading or absolutely worthless, I respectfully decline to make other answers to your enquiries.

Respectfully yours,

H. N. BUTTOLPH, M. D.,

Superintendent of Asylum.

XLIII.

For Ward's Island Asylum see end of Appendix A.

XLIV.

MONROE CO. INSANE ASYLUM,

ROCHESTER, N. Y., December 21, 1883

DEAR SIR: Yours of the 19th at hand. In this small county asylum, no properly tabulated records have been made or kept until a very recent date. Our patients here are all of a chronic class, having been treated in the State Asylum for two years or more before they are sent here for custodial care, and it is not always an easy matter to determine the true causes of the mental malady.

In my opinion, however, alcohol *is not now as operative as a cause of mental disease* as it was ten years ago, or rather there are fewer cases of insanity traceable to alcoholism in this community than was formerly the case.

Very respectfully yours, W. L. LORD,
Superintendent.

XLV.

ERIE COUNTY ASYLUM,

BUFFALO PLAINS, N. Y.

Year.	Number of Patients Admitted.	Number of Cases of Insanity caused by Intemperance. Male.	Female.
1880	334	1	1
1881	353	11	1
1882	339	2	1

XLVI.

HUDSON RIVER STATE HOSPITAL,

POUGHKEEPSIE, N. Y., December 1, 1883.

Year.	Number of Patients Admitted.	Number of Cases of Insanity caused by Intemperance. Male.	Female.
1872	219	10	2
1873	199	10	..
1874	190	8	1
1875	132	8	..
1876	153	5	1
1877	172	5	2
1878	139	3	2
1879	128	5	1
1880	160	6	2
1881	179	12	..
1882	211	13	..
1883	277	22	2

DEAR SIR: This hospital was opened in October, 1871, as an institution for the treatment and cure of cases of insanity. We do not receive inebriates unless they have been pronounced insane, and as such are committed here. We have, therefore, had but few cases of what may be properly termed "alcoholism." The above table shows the number of patients received since the opening of the hospital in which intemperance was the supposed exciting cause of mental derangement as compared with the whole number of admissions.

Very truly yours, S. M. CLEAVELAND.

XLVII.

STATE LUNATIC ASYLUM,

Utica, New York.

Year.	Number of Patients Admitted.	Number of Cases of Insanity caused by Intemperance.	
		MALE.	FEMALE.
1850	367	21	2
1851	366	44	1
1852	390	41	5
1853	424	61	3
1854	390	42	3
1855	275	35	..
1856	242	27	2
1857	235	31	2
1858	333	20	..
1859	312	24	1
1860	337	19	..
1861	295	15	1
1862	287	14	2
1863	287	12	..
1864	319	31	..
1865	356	30	2
1866	388	18	..
1867	401	21	4
1868	382	21	2
1869	463	19	2
1870	481	25	2
1871	516	36	3
1872	399	33	2
1873	410	39	6
1874	368	29	5
1875	432	43	6
1876	436	33	2
1877	460	30	5
1878	427	20	5
1879	418	27	1
1880	468	27	2
1881	411	25	4
1882	412	28	4
1883	404	36	2

The drunkards are not included in the number of cases of alcoholism.

XLVIII.

ST. VINCENT'S REFUGE FOR THE INSANE,

at Harrison, N. Y.

Year.	Number of Patients Admitted.	Number of Cases of Insanity caused by Intemperance.	
		MALE.	FEMALE.
1883	39	..	1

XLIX.

STATE HOMŒOPATHIC ASYLUM FOR THE INSANE,

MIDDLETOWN, Orange Co., N. Y.

Year.	Number of Patients Admitted.	Number of Cases of Insanity caused by Intemperance. MALE.	FEMALE.
1874	69	2	..
1875	152	..	..
1876	195	1	1
1877	228	..	..
1878	284	2	1
1879	283	2	1
1880	311	1	1
1881	340	4	..
1882	391	9	2
1883	509	5	..

L.

MARSHALL INFIRMARY,

TROY, New York.

Year.	Number of Patients Admitted.	Number of Cases of Insanity caused by Intemperance. MALE.	FEMALE.
1854	39	2	..
1855	98	2	1
1856	119	5	1
1857	101	3	3
1858	164	11	2
1859	139	13	3
1860	193	24	5
1861	382	26	..
1862	222	26	5
1863	169	27	..
1864	220	27	2
1865	213	44	2
1866	132	37	1
1867	167	40	3
1868	223	68	3
1869	202	49	1
1870	260	58	4
1871	232	60	1
1872	369	65	3
1873	227	70	4
1874	234	76	4
1875	218	60	3
1876	251	77	6
1877	221	73	3
1878	165	49	6
1879	194	57	
* 1880	419	106	6
* 1881	491	63	4
1882	274	109	5

* Small-pox epidemic.

LI.

BUFFALO STATE ASYLUM FOR THE INSANE.

Year.	Number of Patients Admitted.	Number of Cases of Insanity caused by Intemperance. MALE.	FEMALE.
1881	219	28	6
1882	273	25	7
1883	265	28	7

The asylum was opened for patients in November, 1880. During the year 1881 there were admitted 219 patients. Of this number, intemperance was recognized as a causative element in 34 cases; out of 273 cases in 1882, admitted in 32 cases. In 1883 ; 35 cases out of 265.

Acute cases are received from the eight Judicial districts, and counties of Monroe and Steuben.

LII.

SANFORD HALL, PRIVATE ASYLUM,

Flushing, N. Y.

Year.	Number of Patients Admitted.	Number of Cases of Insanity caused by Intemperance. MALE.	FEMALE.
1854	59	..	..
1855	62	2	..
1856	65	4	..
1857	63	1	1
1858	48	2	..
1859	58	3	..
1860	58	1	..
1861	48	3	..
1862	53	5	1
1863	49	..	1
1864	61	6	..
1865	53	6	..
1866	52	8	..
1867	62	7	1
1868	49	4	2
1869	47	4	..
1870	41	..	..
1871	44	..	..
1872	43	2	..
1873	47	3	1
1874	46	1	..
1875	50	2	..
1876	49	2	1
1877	49	4	1
1878	39	2	..
1879	41	1	..
1880	44	1	..
1881	46	..	..
1882	45	2	1

LIII.

ASYLUM, MINNEOLA, N. Y., December 3, 1883.

DEAR SIR: Yours of the 30th received, in reply would say I have been through the wards of the Queen's County Asylum for the past five years in my official capacity, but have had charge only since the last month, and during that time have had a more careful study of the patients; I can say to your enquiries we have but two cases, one male and one female, attributable to the use of strong drink; we number at this time one hundred and twenty, 58 males and 62 females. Hoping these few lines may meet the desired end and your full approval, I am, etc.,

Respectfully,

CHAS. H. CLEMENT,
Superintendent.

LIV.

INSANE ASYLUM,

ONONDAGA, N. Y., December 28th, 1883.

DEAR SIR:—Your favor of November 30th, reached me in due time, but owing to a total lack of statistics I have neglected to reply to it. No books relating to this subject have been kept in the institution, and I have been here less than six months, so I am hardly in a position to answer it.

However, I should judge that at least one half of the males admitted during my service here have used alcohol to excess, but its share in producing the insanity I am not prepared to state, but in at least four cases it has seemed to me that it was the direct *exciting* cause, but the predisposition existed before. The habits of those who have been here for a long time previous to their coming here are something of which I know nothing.

This report is exceedingly imperfect, but I can do no better in the total absence of records.

Yours respectfully,

A. A. ALDRICH,
Med. Superintendent.

LV.

INSANE ASYLUM OF NORTH CAROLINA,

RALEIGH, N. C., November 19th, 1883.

DEAR SIR:—Your favor of the 15th, asking for certain information relative to "alcoholism" received. Under the laws of committal of insane persons in this State, so little of the previous history is required and the facts elicited are so meagre that it is generally very unsatisfactory and unreliable, and among the least reliable are the *causes* many of which are no doubt collateral, or incidental circumstances and not causes. We have to rely upon opinions given upon little formality, elicited by magistrates of but little experience, and frequently upon observation of no physician at all. My observations, during sixteen years experience here in an institution with a capacity for only 250 patients, does not indicate any very considerable amount of insanity produced by alcoholism. It falls short I think of that produced by excessive use of tobacco and opium. This observation is to be taken however in connection with the fact that cases of dipsomania are not admitted here.

Yours truly,

EUGENE GRISSOM.

LVI.

LUNATIC ASYLUM,

EAST PORTLAND, Oregon.

DEAR SIR:—It is quite impossible from the data I have at hand to furnish you the desired information. I am quite satisfied from personal observations that much insanity is caused by alcoholism, especially by that habit continued over a long period, and resulting in chronic and irreparable injury to the brain.

Yours truly,

S. E. JOSEPHI.

LVII.

DAYTON ASYLUM FOR THE INSANE,

DAYTON, OHIO, December 13, 1883.

DEAR SIR: This asylum was opened September 1, 1855, since which time there have been admitted 2,744 males, 2,665 females. Of this number the cause of insanity alleged to be intemperance for males is 221, females 10. Owing to the great amount of work now on hand and want of time, it is impossible to separate the number into their several years.

Very respectfully,

H. A. COBEY, *Superintendent*,
Per J. A. M., *Sec.*

LVIII.

COLUMBUS ASYLUM FOR THE INSANE.

Year.	Number of Patients Admitted.	Number of Cases of Insanity caused by Intemperance. MALE.	FEMALE.
1881	329	12	1
1883	290	16	..

LIX.

LONGVIEW ASYLUM,

CARTHAGE, OHIO.

Year.	Number of Patients Admitted.	Number of Cases of Insanity caused by Intemperance. MALE.	FEMALE.
1882	4647	398	85

LX.

ATHENS ASYLUM FOR THE INSANE,

ATHENS, OHIO.

Year.	Number of Patients Admitted.	Number of Cases of Insanity caused by Intemperance.	
		MALE.	FEMALE.
1874	701	32	3
1875	252	17	
1876	241	13	..
1877	297	12	1
1878	212	7	1
1879	202	11	..
1880	240	7	..
1881	198	17	1
1882	216	16	..
1883	216	13	3

The above represents the total number in which intemperance was attributed as a cause *directly*; indirectly it was a factor in many more cases, especially among females.

A. B. RICHARDSON,
Superintendent.

LXI.

CLEVELAND ASYLUM FOR THE INSANE,

CLEVELAND, OHIO, December 4, 1883.

DEAR SIR: It has been my intention for some time past to answer your previous communications touching the matter of alcoholism in this asylum. Absence from here, and sickness in my family, have prevented me from so doing, and I trust you will appreciate the force of my reasons for this delay in writing you.

I have to inform you that this asylum was burned in September, 1872, and at that time so many of the records and former reports were destroyed, I cannot give you all the details you desire. The asylum has been in operation twenty-nine years, and during that period 5,532 patients have been admitted, and in the table of causes 280 cases are ascribed to intemperance as a cause of their insanity. During the last year we admitted 244 patients, and in eighteen of these cases the cause was attributed to intemperance. The population from which we draw our cases numbers 700,000, is partly urban and partly rural, but the capacity of our aylsum (640) is sufficient to accommodate more than one-half of the existing, and three-fourths of the occurring cases. The causation of insanity is of a complex character, and I do not see how it will be possible to determine with anything like positiveness what percentage of cases is caused by alcohol alone. Permit me to refer you to a little book, written by Dr. D. Hack Luke in 1878, entitled "Insanity and its Production." See page 99. I may have, after a little time an opportunity to write you further on the subject.

Yours truly,

J. STRONG.

LXII.

STATE HOSPITAL FOR THE INSANE,

DANVILLE, PA., November 17, 1883.

DEAR SIR: In reply to yours of the 15th inst. making inquiry as to the prevalence of mental disturbance from use of alcoholic beverages, I wish to say briefly, that the causes assigned by friends of patients for their insanity are very often very incorrectly given, and that the statistics thus made up are both defective and erroneous, as incidental remarks made long subsequently to the admission of patients, clearly show. I give for this reason no statistics; my opinion is that there is much mental defect and insanity from the effect of alcohol taken by the sufferer himself, and also by the parents of the patient. Within a month, two brothers have been admitted made insane with excessive drink *

Very truly yours, S. S. SCHULTZ,
Superintendent.

* Subsequent inquiry revealed the fact, that nothing reliable bearing on this subject can be obtained.

LXIII.

PHILADELPHIA HOSPITAL,

11 Month, 21, 1883.

DEAR SIR: Aside from heredity, I believe there is no cause so frequently productive of insanity, as the excessive use of alcoholic beverages.

Very respectfully, D. D. RICHARDSON,
Physician-in-Chief.

LXIV.

WESTERN PENNSYLVANIA HOSPITAL FOR THE INSANE,

DIXMONT, ALLEGHENY CO., PA., December 3, 1883.

DEAR SIR: In answer to your letter, I may say that no record is kept in this asylum that would enable us to answer your questions fully and satisfactorily to ourselves, without an immense amount of work.

I think, however, that about one-third of the cases upon the male side of the house are brought here through a disease of brain resulting from the use of alcohol.

Very truly yours, J. A. REED, M. D.

LXV.

STATE HOSPITAL FOR THE INSANE,

NORRISTOWN, PA.

Year.	Number of Patients Admitted.	Number of Cases of Insanity caused by Alcoholism in its widest sense. MALE.	FEMALE.
1880	295	21	..
1881	424	30	..
1882	494	35	..
1883	532	31	..

The female department is entirely separate from the male and in charge of Dr. Alice Bennett. Patients are drawn from Philadelphia and the six Southeastern counties of Pennsylvania.

LXVI.

STATE HOSPITAL FOR THE INSANE,

WARREN, PENN.

Year.	Number of Patients Admitted.	Number of Cases of Insanity caused by Intemperance. MALE.	FEMALE.
1881	199	3	..
1882	355	6	..
1883	423	6	1

LXVII.

STATE LUNATIC HOSPITAL,

HARRISBURG, PA.

Year.	Number of Patients Admitted.	Number of Cases of Insanity caused by Intemperance. MALE.	FEMALE.
1880	121	2	..
1882	64	4	3

From the beginning 4,563 patients of which 125 suffered from the effects of intemperance.

LXVIII.

ASYLUM FOR THE RELIEF OF PERSONS DEPRIVED OF THE USE OF THEIR REASON,

PHILADELPHIA, PA.

Report of 1883 shows intemperance to have been the cause of insanity in 118 cases out of a total of 1,063, admitted during a number of years.

LXIX.

PENNSYLVANIA HOSPITAL FOR THE INSANE.

Number of insane persons admitted to this department of the hospital from the time it was opened in October, 1859, to the first of January, 1884, was 2,212.

Of these 2,212 persons the insanity was caused by intemperance in 285 cases. Of these 285 the following number would sooner or later have become insane from other causes: 56.

Of the insane persons admitted (2,212), 13 per cent.(285) became so from intemperance.

S. PRESTON JONES,
Medical Superintendent.

LXX.

BURN BRAE,

CLIFTON HEIGHTS, Delaware Co., Pa., Dec. 4, 1883.

DEAR SIR:—Regretting inability to fill out your blank, would at the same time say that the result of my observations shows that in a large proportion of cases of general paralysis of insane, the history received with them declares them to have been intemperate previous to the development of the disease.

Very respectfully, R. A. GIVEN,
Per J. H. P.

LXXI.

BUTLER HOSPITAL,

PROVIDENCE, Dec. 15, 1883.

Year.	Number of Patients Admitted.	Number of Cases of Insanity caused by Intemperance.	
		MALE.	FEMALE.
1850	113	4	..
1851	127	3	1
1852	142	5	1
1853	136	2	2
1854	131	1	..
1855	137	..	1
1856	142	2	..
1857	140	..	1
1858	135	2	..
1859	135	3	..
1860	127	3	..
1861	135	4	..
1862	132	3	..
1863	130	3	..
1864	132	6	.
1865	131	1	..
1866	119	9	..
1867	131	11	1
1868	150	5	..
1869	154	6	..
1870	95	5	1
1871	114	12	..
1872	134	16	..
1873	129	10	..
1874	127	7	..
1875	143	8	..
1876	145	7	1
1877	156	12	7
1878	170	8	3
1879	148	11	2
1880	171	14	2
1881	166	14	2
1882	183	11	2
1883	195	7	4

The patients are mainly from Rhode Island, which has a population of a little less than 300,000, but many come from other states.

LXXII.

SOUTH CAROLINA LUNATIC ASYLUM,

COLUMBIA.

Year.	Number of Patients Admitted.	Number of Cases of Insanity caused by Intemperance.	
		MALE.	FEMALE.
1870	322	8	..
1871	205	5	..
1873	309	5	1
1876	300	14	..
1878	306	22	..
1879	375	11	1
1880	420	9	..
1881	490	14	1
1882	550	16	..
1883	603	19	2

LXXIII.

TENNESSEE HOSPITAL FOR THE INSANE.

Year.	Number of Patients Admitted.	Number of Cases of Insanity caused by Intemperance.	
		MALE.	FEMALE.
1870	352	10	1
1871	369	10	1
1872	372	5	5
1873	370	8	8
1874	371	9	9
1875	380	7	4
1876	386	8	3
1877	372	10	4
1878	376	11	5
1879	379	7	2
1880	388	6	4
1881	396	4	4
1882	408	5	2

LXXIV.

VERMONT ASYLUM FOR THE INSANE,

BRATTLEBORO, VT., November 19, 1883.

DEAR SIR:—Your circular letters duly received. In reply would say that I am able to give you the exact percentage of cases due to intemperance in the admission to this Institution for ten years past, during which I have been in charge. I cannot without great labor eliminate them from preceding admissions.

The percentage for the time mentioned is about *six and four tenths* per cent.

Yours truly,

J. DRAPER,

Superintendent.

LXXV.

WESTERN LUNATIC ASYLUM,

STAUNTON, VIRGINIA.

Year.	Number of Patients Admitted.	Number of Cases of Insanity caused by Intemperance. MALE.	FEMALE.
1850	348	..	..
1851	406	..	..
1852	438	..	..
1853	460	..	..
1854	464	..	..
1855	457	..	..
1856	463	..	..
1857	455	..	..
1858	435	29	..
1859	431	29	..
1882	519	29	2
1883	534	32	2

Sorry we havn't the means at hand now to give you all the above statistics.

Very truly, R. S. HAMILTON,
Per H.

LXXVI.

EASTERN LUNATIC ASYLUM,

WILLIAMSBURG, VIRGINIA, November 20, 1883.

Year.	Number of Patients Admitted.	Number of Cases of Insanity caused by Intemperance. MALE.	FEMALE.
1868	242	..	..
1869	224	..	..
1870	273	..	..
1871	265	..	..
1872	288	2	..
1873	311	3	1
1874	234	2	1
1875	366	1	..
1876	357	1	..
1877	378	3	..
1878	376	5	..
1879	367	1	..
1880	408	4	..
1881	426	5	..
1882	469	3	..
1883	536	9	..

Records prior to this time incomplete on this point.

From the beginning (1868), 1208 patients, of which 46 suffered from the effects of intemperance.

RICHARD A. WISE, M. D.,
Superintendent.

LXXVII.

CENTRAL LUNATIC ASYLUM,

RICHMOND, VA.

Year.		Number of Patients Admitted.		Number of Cases of Insanity caused by Intemperance.		
				MALE.		FEMALE.
1882-83		184	....	1	....	..

LXXVIII.

WEST VIRGINIA HOSPITAL FOR THE INSANE,

WESTON, WEST VA.

Year.		Number of Patients Admitted.		Number of Cases of Insanity caused by Intemperance.		
				MALE		FEMALE.
1883	....	685	....	59	... .	..

LXXIX.

WISCONSIN STATE HOSPITAL FOR THE INSANE,

MENDOTA, October 29, 1883.

DEAR SIR: I am sorry to state that our statistical tables for the past twenty-three years will not allow me to fill out your blank. It could not be done except to go over all the cases individually in our books, which would take months to do, and then it would be only guess-work, as the friends and relatives *will not admit alcoholism as a cause* of insanity in these particular cases.

Very respectfully,

R. M. WIGGINGTON,

Superintendent.

LXXX.

MILWAUKEE ASYLUM FOR THE INSANE,

WAUWATOSA, WIS.

Year.		Number of Patients Admitted.		Number of Cases of Insanity due to use of Alcoholic Liquors and also Cases classified as Inebriates.		
				MALE.		FEMALE.
1880	....	245	....	18	....	..
1881	....	126	....	11	....	..
1882	....	96	....	8	....	..
1883	....	117	....	21	....	..

NEW YORK CITY ASYLUM FOR THE INSANE,

WARD'S ISLAND, N. Y.

DEAR SIR: Enclosed please find the tables which extend from 1876 to 1881. * * In considering their probable bearing on the subject of your inquiry, it is necessary to understand that a relatively small number of those given under the heading of Intemperance is *solely* due to that cause, but that in the majority several causes co-operate; whilst it is not denied that intemperance may be at the root of other influences, as loss of work and consequent destitution, etc.,

Very respectfully,

A. E. MACDONALD,
Medical Superintendent.

Showing the habits of persons admitted or discharged respectively at the New York City Insane Asylum, Ward's Island:

Year.	Admitted.	Intemperate.	Temperate.	Abstinent.	Unknown.
1876	381	205	129	24	23
1877	494	225	210	51	8

In 1878 the statistical tables were changed so as to exhibit the habits of those "discharged during the year," the information in this case being much more reliable than in the case of those recently admitted:

YEAR.	ADMITTED.	DISCHARGED.				
		Intemperate.	Temperate.	Abstinent.	Unknown.	Total.
1878	467	127	60	3	55	254
1879	393	139	95	3	49	286
1880	436	181	139	2	60	382
1881	512	197	131	24	107	436

APPENDIX B.

TABLES II, III, IV AND V.

STATISTICS OF INEBRIETY.

TABLE II.

Showing kind and quantity of drinks used by each of 500 Inebriates, treated in the KINGS COUNTY INEBRIATES' HOME at FORT HAMILTON, N. Y.

No.	Age.	Sex.	Place of Birth.	Education.	Age at which Inebriate began using intoxicants.	Kind of drink (fermented or distilled) used.	Maximum daily quantity that patient consumed of such drink.	Occupation when using intoxicants was commenced.	Mental or physical cause to which excesses are ascribed.	Kind of beverages to which necessity of med'l treatm'nt is attributed.	Number of attacks of delirium.
1	27	Male.	U. S.	Com'n School.	25	Distilled.	Largely.	Distiller.	Business.	Distilled.	2
2	50	"	"	Collegiate.	48	"	10 drinks.	Real Estate.	Sexual Excesses	"	
3	46	"	Germany.	Ordinary.	46	"	1 gill.	Groceryman.	Association.	"	
4	37	"	"	"	23	Both.	6 gl. beer & 2 w'ky.	Piano Tuner.	Injury,	"	
5	29	"	U. S.	Com'n School.	16	Distilled.	6 drinks.	Mason.	"	"	
6	47	"	"	Fair.	17	"	1 pint.	Carpenter.	Association.	"	
7	36	"	Ireland.	None.	26	All kinds.	1 "	Laborer.	"	"	
8	27	"	U. S.	Com'n School.	23	Distilled.	½ "	Carpenter.	"	"	
9	36	"	"	Collegiate.	20	"	1 "	Clerk.	"	"	
10	25	"	"	"	21	"	2 drinks to 2 quarts	Clerk.	Disease.	"	
11	32	"	Ireland.	"	22	Beer.	1 quart.	Editor.	Association.	Fermented.	
12	47	"	U. S.	"	27	Distilled.	1 pint.	Painter.	"	Distilled.	2
13	37	"	"	"	29	"	"	Lather.	"	"	
14	27	"	"	"	17	"	"	None.	"	"	
15	36	"	"	"	16	"	"	Newsdealer.	"	"	
16	32	Female.	Ireland.	"	30	"	"	Housewife.	Injury to head.	"	
17	48	"	"	None.	33	Fermented.	2 quarts.	Domestic.	"	Fermented.	
18	29	Male.	U. S.	Com'n School.	20	Distilled.	Moderate.	Dry Goods.	Association.	Distilled.	
19	41	"	"	"	21	"	1½ pint.	Stock Broker.	Business troubles	"	
20	27	"	"	"	17	All kinds.	1 pint.	Clerk.	Association.	"	
21	29	"	"	"	20	Distilled.	"	Shoe Cutter.	Sunstroke.	"	
22	37	"	"	"	21	All kinds.	"	Soldier.	Association.	"	
23	33	"	"	"	23	Distilled.	"	Baker.	"	"	
24	62	"	Ireland.	"	22	"	"	Tailor.	Injury to head.	"	
25	60	"	Scotland.	"	30	"	Largely.	Designer.	Association.	"	1
26	30	"	Ireland.	"	18	"	Plenty.	Laborer.	"	"	
27	39	"	"	.	21	"	½ pint.	Clerk.	"	"	
28	50	"	U. S.	"	32	Both.	....	Merchant.	Loss of Wife.	"	
29	38	"	"	"	30	"	Largely.	Bookkeeper.	Association.	"	1
30	34	"	"	Collegiate.	20	Distilled.	"	Editor.	"	"	
31	41	"	"	"	29	"	1 quart.	Lecturer.	Melancholia.	"	
32	35	"	Canada.	Com'n School.	22	Both.	1 gill liq., ale & beer.	Bank Clerk.	Mental Depres'n.	"	
33	41	"	U. S.	"	16	Distilled.	1 pint.	Manufacturer.	Association.	"	1
34	58	"	England.	"	21	"	"	Merchant.	"	"	1
35	38	"	U. S.	.	21	All kinds.	1 pint distilled.	Mechanic.	Domestic trouble.	"	1

36	45	Male.	U. S.	Com'n School.	20	Distilled.	1 pint.	Broker.	Association.	Distilled.	
37	49	"	England.	Collegiate.	19	"	"	Merchant.	"	"	3
38	42	"	U. S.	Com'n School.	40	All kinds.	Largely.	Real Estate.	"	"	
39	49	"	"	"	24	"	25 drinks.	Hatter.	"	"	
40	38	"	"	"	20	Fermented Beer.	1 quart.	Clerk.	"	Fermented.	
41	45	"	"	"	20	Distilled.	1 pint.	Broker.	"	Distilled.	
42	49	"	England.	"	19	"	"	Engraver.	Injury.	"	
43	23	"	U. S.	Rudimentary.	15	"	"	Hatter.	Association.	"	
44	34	"	Ireland.	None.	22	"	1 quart.	Farmer.	"	"	
45	25	"	"	"	18	"	1 pint.	Laborer.	"	"	
46	46	"	U. S.	"	28	"	1 quart.	"	"	"	2
47	27	"	"	Com'n School.	12	Dist. and Opium.	"	Bookkeeper.	"	"	
48	51	"	"	Collegiate.	21	Dist. (Brandy).	"	Actor.	"	"	4
49	48	"	Scotland.	"	26	Distilled.	1 pint.	Lawyer.	"	"	1
50	33	"	U. S.	Com'n School.	18	"	½ pint.	Hatter.	"	"	1
51	35	"	"	"	20	"	1 quart.	Machinist.	Association.	"	
52	42	"	"	"	22	All kinds.	½ pint distilled.	Carpenter.	"	"	
53	29	"	"	"	18	"	1 qt. lager, 1 pt. w'ky	None.	"	"	
54	26	"	England.	"	20	Ale and Lager.	From 1 qt. to 2 gals.	Artist.	"	Fermented.	
55	22	"	U. S.	Collegiate.	18	Beer only.	15 glasses.	Lawyer.	"	"	
56	24	"	Ireland.	Rudimentary.	14	Distilled.	½ pint.	Livery stable.	"	Distilled.	1
57	26	"	U. S.	"	24	"	1 pint.	Merchant.	"	"	
58	41	"	"	Com'n School.	21	"	"	Clerk.	"	"	
59	45	"	"	"	21	"	"	"	"	"	
60	48	"	"	Collegiate.	14	"	"	Lawyer.	Injury to leg.	"	
61	24	"	"	"	14	"	20 drinks	"	Association.	"	
62	32	"	"	"	22	"	½ pint.	Clerk.	"	"	
63	34	"	"	Com'n School.	19	Whiskey and Ale.	1 pint, besides ale.	Broker.	Disease.	"	1
64	37	"	Ireland.	"	22	Distilled.	1 pint.	Carman.	"	"	1
65	48	Female.	"	None.	28	Ferm'd (beer only)	Largely.	Housekeeper.	Association.	Fermented.	
66	44	Male.	Canada.	Collegiate.	26	Distilled.	½ pint.	Physician.	Insomnia.	Distilled.	1
67	28	"	Ireland.	Com'n School.	20	W'key, Wine, Beer	Largely.	Wine Dealer.	Business.	"	
68	26	"	U. S.	"	20	Distilled.	1 pint.	Conductor.	Association.	"	
69	33	"	"	"	18	"	"	Painter.	"	"	
70	39	Female.	Canada.	"	29	"	½ pint.	Domestic.	"	"	
71	45	Male.	U. S.	Collegiate.	25	"	1 pint.	Lawyer.	Injury to head.	"	
72	46	"	England.	Com'n School.	36	"	"	None.	Association.	"	1
73	38	"	Ireland.	"	20	"	1 quart.	Painter.	"	"	4
74	41	"	U. S.	"	25	"	½ pint.	Clerk.	"	"	
75	33	"	"	"	24	"	1 pint.	Machinist.	"	"	2
76	22	"	"	"	15	"	1 quart.	Painter.	Injury to head.	"	
77	40	"	"	"	20	"	"	Paper Ruler.	Association.	"	4
78	33	"	"	"	18	"	......	Merchant.	"	"	
79	38	"	"	"	13	"	1 quart.	Expressman.	"		
80	42	"	"	"	22	"	Largely.	None.	"		
81	24	"	"	"	17	"	1½ pint.	"	"	"	
82	22	"	"	"	17	"	1 pint.	Butcher.	"	"	
83	32	"	"	"	20	"	1½ quart.	Farmer.	"	"	
84	27	"	"	"	13	"	1 quart	Clerk.	"	"	
85	43	"	"	"	18	"	1 pint.	Advert'g Agt.	Business troubles	"	

TABLE II—Continued.

Showing kind and quantity of drinks used by each of 500 Inebriates, treated in the KINGS COUNTY INEBRIATES' HOME at FORT HAMILTON, N. Y.

No.	Age.	Sex.	Place of Birth.	Education.	Age at which Inebriate began using intoxicants.	Kind of drink (fermented or distilled) used.	Maximum daily quantity that patient consumed of such drinks.	Occupation when using intoxicants was commenced.	Mental or physical cause to which excesses are ascribed.	Kind of beverages to which necessity of med'l treatm'nt is attributed.	Number of attacks of delirium.
86	36	Male.	Ireland.	Rudimentary.	28	Wine and Beer.	1 quart.	Merchant.	Association.	Fermented.	
87	23	Female.	U. S.	Com'n School.	22	Distilled.	1 pint.	None.	"	Distilled.	
88	26	Male.	"	"	16	"	1 quart.	Horseshoer.	"	"	1
89	30	"	Canada.	"	18	Wky. & Absinth.	"	Salesman.	"	"	1
90	49	"	U. S.	"	24	Whiskey and Ale.	1 pt. wky., beer & ale	Farrier.	"	"	
91	33	"	"	"	30	Dist., also Opium.	1 pt. 4 grains opium.	Switchman.	Injury.	"	
92	41	"	Ireland.	"	25	Distilled.	5 to 15 glasses.	Rigger.	"	"	1
93	19	"	U. S.	"	15	All kinds.	15 drinks.	Printer.	Association.	"	
94	28	"	Ireland.	"	18	Distilled.	10 drinks.	Bartender.	Business.	"	1
95	36	"	U. S.	"	27	Gin and Ale.	"	Manufacturer.	Association.	"	1
96	37	"	"	"	21	Distilled.	1 pint.	None.	"	"	
97	25	"	Ireland.	"	18	"	20 drinks.	Bookkeeper.	Business.	"	1
98	42	"	U. S.	"	21	"	"	"	Family troubles.	"	1
99	18	"	"	"	17	Beer & Whiskey.	20 to 30 drinks.	Clerk.	Association.	Both.	
100	43	"	Ireland.	"	18	Distilled.	10 to 40 drinks.	Tinsmith.	"	Distilled.	1
101	32	"	U. S.	"	22	"	10 drinks.	Mechanic.	"	"	
102	24	"	Cuba.	"	22	"	5 drinks.	Clerk.	Mental depres'n.	"	
103	64	"	Ireland.	"	54	"	10 drinks.	Tailor.	Association.	"	
104	38	"	U. S.	"	20	"	20 to 30 drinks.	Salesman.	"	"	1
105	41	Female.	Ireland.	"	35	Whiskey & Beer.	1 pint.	None.	Disease g'l organs	"	
106	28	Male.	U. S.	"	16	Distilled.	10–30 drinks.	Broker.	Bullet Wound.	"	1
107	64	"	"	"	34	Dist. Liq. & Ale.	10–50 drinks.	"	Domestic trouble.	Both.	
108	59	"	Ireland.	"	19	Whiskey and Ale.	10 drinks.	Saddler.	Wound.	"	
109	38	"	U. S.	"	18	All kinds.	30 drinks.	Clerk.	Injury to Head.	Distilled.	
110	32	"	"	"	16	Whiskey and Ale.	12 drinks.	Bookkeeper.	Association.	"	
111	32	"	Ireland.	"	30	Whiskey & Beer.	8 dr'ks wh'ky & b'r.	Clerk.	"	"	
112	26	Female.	"	"	19	Dist. and Opium.	1 pint.	Nurse.	Disease.	"	
113	46	Male.	U. S.	"	21	Both.	½ pint whiskey.	Mechanic.	Association.	"	
114	24	"	"	"	22	Distilled.	5 glasses.	Engin'r in dist.	Business.	"	
115	26	"	"	"	17	Whiskey & Beer.	8 glasses whiskey.	Restaurant.	Business trouble.	"	
116	41	"	"	"	39	"	10 drinks.	Coal Dealer.	Trouble.	Both.	
117	61	"	England.	"	49	Distilled.	10–20 glasses.	Shoemaker.	Injury to Head.	Distilled.	
118	45	Female.	Ireland.	"	30	"	½ pint.	None.	Association.	"	
119	33	"	U. S.	"	28	"	1 pint.	"	Family Trouble.	"	1
120	40	Male.	"	Collegiate.	35	"	1 pint.	Lawyer.	Association.	"	

121	30	Male.	U. S.	Com'n School.	13	Whiskey & Beer.	20 drinks.	Plumber.	Association.	Both.	
122	55	Female.	"	"	51	Distilled.	1 pint.	None.	"	Distilled.	
123	37	Male.	Ireland.	"	22	Whiskey and Ale.	1 quart.	Liquor Busin's.	Busin's necessity.	"	
124	29	"	U. S.	"	17	Whiskey & Beer.	10 drinks whiskey.	Marble Cutter.	Association.	"	1
125	27	"	"	Collegiate.	12	All kinds.	25 drinks.	Banker.	Trouble.	"	7
126	33	"	Ireland.	Com'n School.	23	Whiskey & Beer.	6 dr'ks dist.&20 b'r.	Cigar Dealer.	Association.	Both.	
127	45	"	U. S.	"	33	Distilled.	7 drinks.	Laborer.	Nervous Temp't.	Distilled.	
128	29	"	Ireland.	"	27	Brandy and Beer.	1 pt.br,sev'l gl.brdy.	None.	Association.	"	
129	31	Female.	Canada.	"	25	Distilled.	1 quart.	Domestic.	"	"	1
130	53	Male.	Ireland.	"	47	"	6 drinks.	None.	Trouble.	"	
131	25	"	U. S.	Collegiate.	19	"	20 drinks.	Lawyer.	Association.	"	
132	37	"	"	Com'n School.	26	Br'dy, Wh'ky, B'r.	1 pint distilled.	Insurance.	Business trouble.		1
133	33	"	England.	"	22	Distilled.	1 pint.	Photographer.	Association.	"	
134	34	"	Ireland.	Rudimentary.	24	"	One quart.	Stone Cutter.	"	"	6
135	40	"	"	Com'n School.	20	Whiskey & Beer	10 drinks w'ky & ale	Bartender.	Business.	"	
136	33	"	U. S.	"	30	Distilled.	1 quart.	Cotton Broker	Association.	"	
137	58	"	"	"	28	"	"	Carpenter.	"	"	
138	33	"	England.	"	15	"	"	Butcher.	Disease.	"	
139	35	"	U. S.	"	25	Ale and Whiskey.	"	Laborer.	"	"	1
140	46	"	"	"	21	Distilled.	20 drinks.	Painter.	"	"	1
141	45	"	"	"	20	W'key, Ale, Beer.	1 qt. sev'l gl. dist'd.	Telegraph Op'r.	Com. of Diseases.	Both.	1
142	23	"	"	"	23	Distilled.	1 quart.	Engineer.	Association.	Distilled.	
143	50	"	"	"	18	Whiskey & Beer	Largely of both.	Printer.	"	Both.	
144	24	"	Germany.	Collegiate.	24	"	1 pt. wky, b'r, wine	Superintend'nt.	"	Distilled.	1
145	58	"	England.	Com'n School.	28	"	½ pt. wky. 1 qt. beer	Druggist.	"	"	
146	53	"	"	"	23	Distilled.	1 pint.	Painter.	"	"	
147	23	Female.	U. S.	"	20	Beer.	2 quarts.	None.	"	Fermented.	
148	58	Male.	"	"	21	Both.	1 pint brandy.	Stationer.	"	Distilled.	1
149	41	"	"	Collegiate.	30	"	½ pt. dis. 2 qts. fer'd	Teacher.	Nervous Temp't.	Both.	
150	43	"	"	Com'n School.	18	Distilled.	2 quarts.	Clerk.	Association.	Distilled.	
151	50	"	"	"	25	Both.	1 pt. dist. b'sides ale	Merchant.	"	"	
152	60	"	England.	"	30	"	" "	Painter.	"	"	
153	36	"	U. S.	"	18	Distilled.	1 quart.	Hatter.	"	"	3
154	36	Female.	Ireland.	"	34	"	1 pint.	Domestic.	"	"	
155	35	Male.	U. S.	"	15	Whiskey & Beer	2 qts. whiskey.	Machinist.	"	"	1
156	51	Female.	England.	"	39	Distilled.	1 pint.	None.	Domestic trouble.	"	
157	38	Male.	U. S.	"	26	Whiskey & Beer	2 qts. br, s'l gl. wky	Bookkeeper.	Association.	"	
158	36	"	"	"	21	Distilled.	1 quart.	Printer.	Shot Wound.	"	
159	60	"	.	Collegiate.	45	"	1 "	Lawyer.	Association.	"	
160	30	Female.	Germany.	Com'n School.	30	Fermented.	3 quarts beer.	Bartender.	Business.	Fermented.	1
161	57	Male.	U. S.	"	27	Distilled.	1 quart.	Clerk.	Association.	Distilled.	1
162	63	"	"	"	43	"	1 quart.	Salesman.	Association.	"	
163	64	"	England.	"	44	Both.	1 pt. wky. beside b'r	Packer.	Trouble.	"	
164	65	"	Scotland.	"	20	Distilled.	1 quart.	Tailor.	Association.	"	1
165	23	"	U. S.	"	20	Both.	3 qts. b'r, 2 glas. w'y	Plumber.	Injury to head.	Both.	
166	30	"	"	"	30	Distilled.	1 quart.	Salesman.	Association.	Distilled.	
167	35	Female.	Ireland.	"	27	Both.	½ pt. b'ndy, 1 pt. b'r	Housewife.	Domestic trouble	"	
168	35	Male.	"	"	15	Distilled.	2 quarts.	Clerk.	Association.	"	
169	25	"	U. S.	"	15	Milk Punch.	1 quart.	Painter.	"	"	
170	34	"	Ireland.	Collegiate.	19	Distilled.	"	Journalist.	"	"	

TABLE II—Continued.

Showing kind and quantity of drinks used by each of 500 Inebriates, treated in the KINGS COUNTY INEBRIATES' HOME at FORT HAMILTON, N. Y

No.	Age.	Sex.	Place of Birth.	Education.	Age at which Inebriate began using intoxicants.	Kind of drink (fermented or distilled) used.	Maximum daily quantity that patient consumed of such drinks.	Occupation when using intoxicants was commenced.	Mental or physical cause to which excesses are ascribed.	Kind of beverages to which necessity of med'l treatm'nt is attributed.	Number of attacks of delirium.
171	48	Female.	U. S.	Com'n School.	20	Both.	1 pt. whiskey & ale	Liquor busin's	Business.	Distilled.	sev'rl
172	26	Male.	"	"	6	Distilled.	1 quart.	Clerk.	Association.	"	
173	57	"	"	"	22	Both.	1 qt. wky., ale or b'r	Printer.	Trouble.	"	
174	43	"	"	"	23	Distilled.	2 quarts.	Cigar Dealer.	Association.	"	2
175	56	"	"	"	36	Both.	1 qt. gin or whiskey	Salesman.	Trouble.	"	6
176	30	"	"	"	19	Distilled.	1 pint.	"	Association.	"	
177	30	"	"	"	17	"	1 quart.	Agent.	"	"	2
178	44	Female.	Canada.	"	34	Both.	1 pt. whiskey & b'r	Housewife.	"	"	
179	41	Male.	U. S.	"	29	All kinds & Op'm	Largely.	Printer.	Hereditary.	Both.	1
180	40	"	England.	"	15	Distilled.	1 quart.	Bookkeeper.	Association.	Distilled.	1
181	39	"	U. S.	"	29	"	1 pint.	Actor.	"	"	
182	35	"	"	"	15	"	1 quart.	Merchant.	Taste acquired.	"	1
183	35	Female.	Ireland,	"	26	"	1 pint.	Housewife.	Association.		1
184	43	Male.	"	None.	26	"	1 quart.	Blacksmith.	"	"	
185	20	"	U. S.	Com'n School.	17	Fermented.	4 quarts.	Clerk.	Injury to head.	Fermented.	
186	28	"	"	"	15	Distilled.	1¼ pint.	Printer.	Association.	Distilled.	1
187	29	"	England.	"	16	Both.	¼ pt. wky., 1 qt. ale	Druggist.	"	Both.	
188	40	"	Ireland.	"	30	"	1 pt. whiskey & b'r	Laborer.	"	Distilled.	
189	35	"	U. S.	Collegiate.	33	Distilled.	1 quart.	Civil Engineer	Habit.	"	
190	34	"	"	Com'n School.	22	"	1 pint.	Bookkeeper.	Association.	"	
191	38	"	"	"	33	Both.	Largely.	Milkman.	Domestic trouble.	Both.	
192	25	"	England.	"	24	"	Unknown.	Clerk.	Injury to head.	Distilled.	
193	32	"	Ireland.	"	25	Distilled.	1 quart.	"	Trouble.	"	
194	32	"	"	"	18	"	1 pint.	Bookkeeper.	Association.	"	1
195	23	"	U. S.	"	18	Fermented.	1 quart to 1 gallon.	Clerk.	"	Fermented.	
196	53	"	Ireland.	"	23	Distilled.	1 pint.	Merchant.	Domestic trouble.	Distilled.	
197	47	"	U. S.	"	22	"	3 pints.	Distiller.	Business.	"	1
198	26		Ireland.	"	20	Both.	1 qt. wky., 1 qt beer	Carpenter.	Injury.	"	2
199	48	Female.	"	None.	33	"	2 gills whiskey.	Housewife.	Trouble.	"	1
200	35	Male.	U. S.	Com'n School.	25	"	Periodical debauch.	Farmer.	Association.	"	
201	55	"	"	"	49	Distilled.	1 quart.	Painter.	"	"	
202	30	"	"	"	30	"	1 pint.	Shoemaker.	Domestic trouble.	"	
203	40	"	"	"	30	"	1 quart.	None.	Diseased condit'n	"	
204	32	"	"	"	17	"	"	Clerk.	Hereditary.	"	
205	44	"	Ireland.	"	20	"	"	Barkeeper.	Business.	"	

206	38	Female.	U. S.	Rudimentary.	25	Both.	3 gills wky., beer.	Servant.	General debility.	Distilled.	
207	29	Male.	"	Com'n School.	22	Fermented.	1 quart to 1 gallon.	Carpenter.	Association.	Fermented.	
208	45	Female.	Ireland.	Rudimentary.	36	Both.	1 qt. distil. at times.	Servant.	Injury to head.	Distilled.	2
209	26	Male.	U. S.	Collegiate.	18	Distilled.	1 quart.	Lawyer.	Association.	"	1
210	38	"	"	Com'n School.	13	"	"	Hotelkeeper.	Business.	"	
211	60	Female.	Ireland.	"	44	"	½ pint.	Servant.	Association.	"	
212	24	Male.	U. S.	"	15	Both.	1 pint distilled.	Junkman.	"		1
213	35	"	"	"	15	Distilled.	1 quart.	Salesman.	"	"	
214	32	"	"	"	16	"	"	Glazier.	"	"	
215	33	"	"	"	15	"	1 to 3 quarts.	Merchant.	"	"	6
216	30	"	"	"	15	Both.	Largely of both.	Bookkeeper.	"	Both.	
217	28	Female.	England.		25	"	1 qt. ale, some wky.	Servant.	Gen'l bad health.	"	
218	31	Male.	Canada.	Com'n School.	24	Both.	1 quart distilled.	Bookkeeper.	Association.	Distilled.	1
219	47	"	Ireland.	None.	35	Distilled.	1 quart.	Carman.	Trouble.	"	3
220	45	Female.	"	"	37	Fermented.	2 quarts.	Servant.	"	Fermented.	
221	38	"	"	Rudimentary.	30	Both.	1 pt. dis., beer, ale.	Housewife.	"	Distilled.	
222	30	Male.	U. S.	Com'n School.	20	Distilled.	2 quarts sometimes.	Clerk.	Association.	"	2
223	50	"	"	"	30	Both.	Largely.	"	"	Both.	
224	47	Female.	Ireland.	None.	36	Distilled.	1 pint.	Housewife.	Domestic trouble.	Distilled.	
225	30	Male.	U. S.	Com'n School.	15	"	1¼ pint.	Clerk.	Association.	"	1
226	55	"	Scotland.	Collegiate.	45	Both.	1 pt. gin, some b'r.	Clergyman.	Disease.	"	
227	38	"	U. S.	Com'n School.	20	Distilled.	1 pint.	Farmer.	Association.	"	
228	33	"	"	"	29	"	"	Bookkeeper.	"	"	
229	42	Female.	Ireland.	Rudimentary.	33	"	"	Servant.	Trouble.	"	
230	25	Male.	U. S.	Com'n School.	20	"	1 quart.	Clerk.	Association.	"	1
231	38	"	"	"	20	"	1 pint to 2 quarts.	Salesman.	"	"	1
232	42	Female.	Canada.	"	39	Distilled.	1 pint.	Housewife.	Trouble at home.	"	
233	48	Male.	"	"	33	Both.	1 pt. whiskey & b'r.	Painter.	Association.	"	
234	39	"	U. S.	"	24	"	1 quart distilled.	Clerk.	Injury to Head.	"	
235	32	Female.	Ireland.	"	26	Both.	1 pint whiskey.	Housewife.	Effects of Misc'ge	"	
236	39	Male.	Canada.	Collegiate.	31	Distilled.	1 pint.	Agent.	Gen'l poor health.	"	
237	46	"	England.	Com'n School.	22	Both.	6 d'ks dist., 1 gal. ale	Carpenter.	Association.	Both.	
238	41	"	U. S.	"	38	Distilled.	1 quart.	Plasterer.	"	Distilled.	1
239	28	"	"	"	21	"	"	Bookkeeper.	"	"	
240	35	"	"	Collegiate.	17	Dist. and Opium.	"	Physician	"	"	
241	34	"	"	"	18	Distilled.	"	Teleg'ph oper'r	Dyspepsia.	"	
242	46	"	Ireland.	None.	44	"	½ pint.	Servant.	Association.	"	
243	32	"	Canada.	Com'n School.	22	"	1 pint.	R. R. Agent.	"	"	
244	41	"	U. S.	"	34	"	1 quart.	Hotel Clerk.	Bus. Trouble.	"	
245	40	Female.	Ireland.	None.	33	Both.	1 pint of each.	Housewife.	Association.	Both.	
246	26	Male.	U. S.	Com'n School.	19	Distilled.	1 pint.	Clerk.	"	Distilled.	
247	47	"	Canada.	"	27	"	"	Machinist.	"	"	
248	56	Female.	Ireland.	None.	46	"	"	Housewife.	Trouble.	"	
249	32	Male.	"	Com'n School.	14	"	1 quart.	Liquor busin's.	Business.	"	
250	32	Female.	"	None.	16	Fermented.	2 quarts.	Housewife.	Disease g'l organs	Fermented.	
251	28	Male.	U. S.	Com'n School.	20	Distilled.	1 quart.	Merchant.	Association.	Distilled.	6
252	35	Female.	"	"	25	"	1 pint.	Housewife.	Trouble.	"	
253	48	Male.	Ireland.	None.	24	"	1½ pint.	Bricklayer.	Association.	"	
254	30	"	U. S.	Com'n School.	15	"	Probably 2 quarts.	Liquor Dealer.	Business.	"	
255	21	"	"	Collegiate.	17	Both.	Largely.	Student.	Association.	Both.	

TABLE II—Continued.

Showing kind and quantity of drinks used by each of 500 Inebriates, treated in the Kings County Inebriates' Home at Fort Hamilton, N. Y.

No.	Age.	Sex.	Place of Birth.	Education.	Age at which Inebriate began using intoxicants.	Kind of drink (fermented or distilled) used.	Maximum daily quantity that patient consumed of such drink.	Occupation when using intoxicants was commenced.	Mental or physical cause to which excesses are ascribed.	Kind of beverages to which necessity of med'l treatm'nt is attributed.	Number of attacks of delirium.
256	34	Female.	Ireland.	None.	18	Distilled.	1 pint.	Housewife.	Domestic trouble.	Distilled.	
257	44	"	"	Com'n School.	26	Both.	½ pt. brandy & beer.	Servant.	Trouble.	"	4
258	46	"	England.	Poor.	26	"	1 pt. brandy & beer.	Housewife.	Domestic trouble.	"	
259	45	"	U. S.	Com'n School.	42	Distilled.	½ pint.	Servant.	Association.	"	
260	58	Male.	"	"	52	"	"	None.	Domestic trouble.	"	
261	39	Female.	"	Rudimentary.	35	"	Largely.	"	Association.	"	
262	41	Male.	"	Com'n School.	25	"	1 quart.	Merchant.	"	"	1
263	40	Female.	Ireland.	None.	32	"	½ pint.	Housewife.	Domestic trouble.	"	
264	34	"	"	Com'n School.	20	"	1 quart.	"	Dom'c tr'ble & inj.	"	
265	50	"	"	Rudimentary.	43	"	½ pint.	None.	Trouble.	"	
266	25	Male.	U. S.	Com'n School.	19	Both.	Largely.	Clerk.	Association.	Both.	
267	30	"	"	Collegiate.	27	Distilled.	Unknown.	"	"	Distilled.	
268	47	"	"	Com'n School.	29	"	1 quart.	Purser.	"	"	
269	40	"	"	Collegiate.	30	"	2 quarts.	Merchant.	"	"	1
270	34	"	Germany.	Com'n School.	14	Both.	1 qt. dist., and beer.	Banker.	"	Both.	2
271	29	"	U. S.	Collegiate.	19	"	2 quarts in all.	Editor.	"	"	1
272	52	"	"	Com'n School.	22	"	1 quart whiskey.	Hatter.	Hereditary.	Distilled.	
273	37	"	"	"	25	"	3 qts. b'r, some wky	Salesman.	Busin's necessity.	Both.	
274	64	Male.	"	Com'n School.	24	Distilled.	1 quart.	Carpenter.	Domestic trouble.	Distilled.	
275	30	"	"	"	22	Both.	1 qt. dist'ld & beer	Clerk.	Association.	"	
276	44	"	"	Collegiate.	24	"	Largely.	Physician.	Domestic trouble.	"	
277	24	"	"	Com'n School.	18	Distilled.	1 quart.	Lard buyer.	Association.	"	
278	43	"	"	"	23	Both.	Largely.	Farmer.	"	"	
279	42	Female.	"	"	30	Distilled.	1 quart.	Dressmaker.	Domestic trouble.	"	
280	23	Male.	"	"	20	"	"	Electrotyper.	Association.	"	
281	60	"	"	"	20	"	30 drinks.	None.	Business matters.	"	
282	21	"	Canada.	"	14	"	20 drinks.	Marble Cutter.	Association.	"	1
283	22	"	Ireland.	"	22	"	1 quart.	Merchant.	"	"	1
284	27	"	U. S.	Collegiate.	17	Both.	½ pt. brandy & beer	Reporter.	"	"	1
285	38	"	"	Com'n School.	18	Distilled.	1 quart.	Stock Broker.	"	"	
286	40	"	"	"	20	"	"	Bookkeeper.	"	"	
287	42	"	"	"	23	"	1½ quart.	Merchant.	Trouble.	"	1
288	30	"	Ireland.	"	15	"	"	Bartender.	Business.	"	1
289	27	"	U. S.	"	17	"	"	Tailor.	Association.	"	
290	38	Female.	Ireland.	"	28	"	1 quart.	Servant.	"	"	1

291	50	Male.	U. S.	Com'n School.	30	Distilled.	1 quart.	Printer.	Association.	Distilled.	2
292	54	Female.	"	"	44	"	1¼ pint.	Milliner.	Bad Health.	"	
293	27	Male.	"	Rudimentary.	27	"	1 quart, or more.	Glass Blower.	Association.	"	10
294	38	"	"	Com'n School.	19	Both.	Largely.	Jeweller.	Habit.	"	1
295	33	Female.	Cuba.	"	30	Distilled	1 quart.	Music Teacher	Domestic trouble.	"	
296	23	Male.	U. S.	Collegiate	20	"	1 pint.	Lawyer.	Association.	"	
297	45	"	"	Com'n School.	30	"	1 quart.	Merchant.	Business.	"	
298	46	Female.	Ireland.	None.	23	"	1 pint.	Cook.	Association.	"	
299	50	Male.	U. S.	Com'n School.	20	"	Largely.	None.	Hereditary.	"	
300	30	"	Ireland.	None.	15	"	1 quart.	Longshoreman	Association.	"	
301	40	Female.	"	"	19	Both.	1 pint whiskey.	Servant.	"	"	
302	37	Male.	U. S.	Com'n School.	18	Distilled.	1 quart.	Druggist.	Wound.	"	
303	32	"	"	"	20	"	"	Peddler.	Trouble.	"	1
304	38	"	"	"	18	"	2 quarts.	Bottler.	Injury to head.	"	
305	52	"	"	"	24	"	1½ quarts.	Farmer.	Disease.	"	
306	40	"	"	"	24	"	1 quart.	Accountant.	Domestic trouble.	"	
307	33	"	Ireland.	"	26	Both.	Ale & 1 qt. dist. liq.	Bartender.	Busin. necessity.	"	1
308	35	"	"	"	15	"	1 qt. of brandy & ale.	"	"	"	2
309	42	"	"	"	32	Distilled.	1 quart.	Servant.	Trouble.	"	
310	27	"	U. S.	"	22	"	1 "	Clerk.	Association.	"	
311	47	"	"	Collegiate.	23	Fermented.	1 gallon.	Merchant.	Injury.	Fermented.	1
312	23	Female.	"	Com'n School.	16	Both.	1 qt. beer & some w.	Housewife.	Insomnia.	Both.	
313	40	Male.	Scotland.	"	23	"	4 qts. " " w'ky.	Printer.	Association.	"	
314	54	"	"	"	24	"	1 qt. w'ky bes. beer.	Gardener.	"	Distilled.	
315	38	"	Ireland.	Collegiate.	34	Distilled.	1 quart.	Dry Goods.	Very nervous.	"	
316	35	"	U. S.	"	33	Dist. also Morph.	Small.	Physician.	Sunstroke.	"	
317	32	"	Ireland.	Com'n School.	22	Distilled.	1½ quart.	Liquor Dealer.	Business.	"	2
318	40	"	U. S.	"	20	"	1 to 3 quarts.	Barkeeper.	"	"	1
319	32	Female.	Ireland.	"	28	"	1 pint.	Housewife.	Trouble.	"	
320	40	"	U. S.	None.	33	Fermented.	3 quarts.	Servant.	Association.	Fermented.	
321	65	Male.	Ireland.	"	25	Distilled.	1½ pint.	Laborer.	"	Distilled.	
322	61	"	U. S.	Com'n School.	21	"	1 quart.	Trainer.	"	"	
323	20	"	"	"	15	"	1 pint.	Salesman.	Sunstroke.	"	
324	48	"	"	Collegiate	28	Both.	Largely.	Merchant.	Injury to head.	"	
325	37	"	Ireland.	Com'n School.	15	Distilled.	1 quart.	Baker.	Association.	"	1
326	39	"	U. S.	"	21	Both.	Largely.	Plumber.	"	"	1
327	40	"	"	"	23	Distilled.	1 to 3 quarts.	None.	"	"	3
328	59	"	"	"	25	Fermented.	3 quarts.	Manufacturer.	Hereditary.	Fermented.	
329	53	"	Germany.	"	21	"	Largely.	Broker.	Association.	"	
330	34	Male.	U. S.	Collegiate.	34	Both.	2 quarts.	Editor.	Sunstroke.	Distilled.	
331	25	"	"	"	18	"	Largely.	Agent.	Father an ineb'te.	"	
332	33	"	"	"	27	Distilled.	1½ quart.	Conductor.	Association.	"	2
333	49	Female.	Ireland.	None.	39	"	1 pint.	Housewife.	"	"	
334	43	Male.	U. S.	Collegiate.	20	"	"	Physician.	"	"	
335	33	"	"	"	20	"	1 to 3 quarts.	Sea Captain.	"	"	1
336	58	"	"	"	28	Both.	Largely.	Lawyer.	Business troubles	"	
337	29	"	"	Com'n School.	17	Distilled.	1 quart.	Clerk.	Association.	"	
338	34	"	"	"	18	"	2 quarts.	"	"	"	
339	29	"	"	"	19	"	Largely.	Bookkeeper.	"	"	
340	31	Female.	Ireland.	None.	.	"	"	Servant.	"	"	1

TABLE II—Continued.

Showing kind and quantity of drinks used by each of 500 Inebriates, treated in the KINGS COUNTY INEBRIATES' HOME at FORT HAMILTON, N. Y.

No.	Age.	Sex.	Place of Birth.	Education.	Age at which Inebriate began using intoxicants.	Kind of drink (fermented or distilled) used.	Maximum daily quantity that patient consumed of such drinks.	Occupation when using intoxicants was commenced.	Mental or physical cause to which excesses are ascribed.	Kind of beverages to which necessity of med'l treatm'nt is attributed.	Number of attacks of delirium.
341	61	Male.	U. S.	Collegiate.	20	Distilled.	1 bottle.	Merchant.	Association.	Distilled.	
342	38	"	"	Com'n School.	20	"	1 quart.	None.	"	"	
343	38	"	"	"	22	"	"	"	"	"	
344	27	"	"	"	20	"	Periodical debauch.	Bookkeeper.	"	"	1
345	36	Female.	Canada.	"	33	"	....	Seamstress.	"	"	
346	50	"	U. S.	None.	49	Fermented.	2 quarts.	Housewife.	"	Fermented.	
347	40	"	England.	"	33	Distilled.	1 quart.	Servant.	"	Distilled.	
348	22	Male.	U. S.	"	12	"	1 pint.	Salesman.	"	"	
349	41	"	"	Collegiate.	36	"	1 bottle.	None.	"	"	2
350	31	"	"	"	26	"	1 pint.	Clerk.	"	"	
351	35	"	"	"	33	"	"	Bookkeeper.	"		
352	49	"	"	"	25	Both.	12 drinks gin & beer	Broker.	"	Both.	
353	46	"	"	Com'n School.	31	Distilled.	1¼ quart.	Merchant.	"	Distilled.	
354	39	"	"	"	30	"	1 quart.	Mechanic.	"	"	
355	24	"	"	None.	18	"	Largely.	Junkman.	"	"	2
356	56	Female.	"	Com'n School.	49	Fermented.	2 quarts.	Housewife.	Domestic trouble.	Fermented.	
357	40	Male.	"	"	25	Distilled.	1 to 3 quarts.	Salesman.	Association.	Distilled.	
358	29	"	Canada.	Collegiate.	26	Both.	1 qt. w'ky, also b'r.	Salesman.	Busin's and asso.	"	1
359	23	"	U. S.	"	21	"	10 drinks.	Bookkeeper.	Association.	Both.	1
360	20		"	Com'n School.	16	Distilled.	1 pint.	None.	Hereditary.	Distilled.	
361	44	"	Germany.	Collegiate.	29		1¼ pints.	Waiter.	Business.	"	
362	50	"	U. S.	Com'n School.	40	"	Largely.	Merchant.	Association.	"	1
363	30	"	"	"	20	"	15 drinks.	Bookkeeper.	"	"	
364	58	"	England.	"	48	"	1 pint.	Engraver.	"		
365	59	"	U. S.	"	49	"	30 drinks.	Builder.	"	"	
366	39	"	Ireland.	Collegiate.	34	"	15 drinks.	Reporter.	"	"	
367	43	"	U. S.	"	33	Both.	Largely.	Merchant.	"	"	
368	37		England.	Com'n School.	20	Distilled.	1 quart.	Carpenter.	"	"	
369	33	"	"	"	23	Both.	1 qt. w'ky, 3 qts. br.	Ironworker.	"	"	
370	37	"	U. S.	"	34	Distilled.	1 quart.	Restaurant.	Business.	"	3
371	36	"	"	Collegiate.	35	"	Largely.	Lawyer.	Mental overext'n.	"	
372	27	"	"	Com'n School.	18	.	"	Bookkeeper.	Association.	"	
373	39	"	"	"	37	"	1 quart.	Agent.	Family troubles.	"	
374	27	"	"	"	17	"	"	M'chant Tailor.	Association.	"	
375	38	"	"	Collegiate.	34	"	30 drinks.	Teacher.	Trouble.		

376	32	Male.	U. S.	Com'n School.	22	Distilled.	1 quart.	Peddler.	Domestic trouble.	Distilled.	3
377	44	"	"	"	24	"	Largely.	Salesman.	Business.	"	
378	37	"	"	Collegiate.	35	"	1 pint.	Journalist.	Association.	"	
379	19	"	"	Com'n School.	16	"	25 drinks.	Bartender.	Father an ineb'te.	"	
380	33	"	Germany.	Collegiate.	30	"	2 quarts.	Druggist.	Family trouble.	"	
381	29	"	U. S.	Com'n School.	19	"	1 quart.	Clerk.	Association.	"	
382	42	"	"	"	22	Both.	12 drinks.	Merchant.	Hereditary.	"	1
383	23	"	"	Collegiate.	22	Fermented.	2 quarts.	Hatter.	Nervous temp't.	Fermented.	
384	28	"	Ireland.	Com'n School.	20	Both.	15 d'ks w., 1 gal. br.	Salesman.	Nervous shock.	Both.	
385	28	"	U. S.	"	18	"	4 d'ks w'y, also b'r.	Carpenter.	Gastritis.	"	1
386	40	Female.	Ireland.	"	37	"	1 pt. whiskey—ale	None.	Genital disorder.	Distilled.	1
387	28	Male.	U. S.	"	22	"	Unknown.	Plasterer.	Association.	"	1
388	49	"	"	Collegiate.	29	Distilled.	20 drinks.	None.	Hereditary.	"	1
389	39	"	Germany.	Com'n School.	15	"	15 drinks.	Butcher.	Association.	"	1
390	28	"	U. S.	"	15	"	Largely.	Clerk.	"	"	1
391	33	"	"	"	31	"	1 pint.	Jockey.	Misfortune.	"	1
392	31	"	"		19	"	1¼ pint.	Roofer.	Hereditary.	"	
393	35	"	Germany.	Collegiate.	30	Both.	1 pint whiskey.	Druggist.	Overwork.	"	
394	34	"	U. S.	Com'n School.	24	Distilled.	½ pint.	Grocer.	Injury to head.	"	1
395	43	"	Germany.	"	18	"	1¼ qt. distilled liq.	Saloon Keeper.	Business.	"	1
396	34	"	Scotland.	Collegiate.	18	All kinds Distil'd.	1 pint.	Journalist.	Association.	"	
397	29	"	U. S.	Com'n School.	19	"	½ pint.	Carpenter.	"	"	1
398	54	"	"	"	14	"	1½ pint.	Pencil-case m'r.	"	"	1
399	31	"	Germany.	"	27	"	1 pint.	Bookkeeper.	Inherited.	"	1
400	60	"	Ireland.	"	54	"	1¼ pint.	Compositor.	Domestic trouble.	"	2
401	22	"	U. S.	"	15	"	15 drinks.	Clerk.	Association.	"	1
402	40	Female.	Ireland.	"	39	Distilled.	1 pint.	None.	Death of mother.	"	
403	28	Male.	U. S.	None.	25	"	8 drinks.	Butcher.	Association.	"	2
404	30	Female.	"	Poor.	27	"	1 pint.	Servant.	"	"	
405	56	Male.	"	Com'n School.	50	"	½ pint.	Carpenter.	Domestic trouble	"	1
406	57	"	"	"	27	"	1¼ pint.	Horse Dealer.	Association.	"	
407	32	"	"	"	27	"	1 quart.	Conductor.	"	"	
408	43	"	"	High School.	18	"	1½ quart.	Broker.	"	"	
409	33	"	"	Com'n School.	21	"	1 pint.	Clerk.	Hereditary.	"	2
410	61	"	"	"	21	'Both.	Largely.	Ferry Master.	Association.	Both.	
411	62	"	"	Collegiate.	42	Distilled.	1 pint.	Lawyer.	Hereditary.	Distilled.	
412	41	"	Ireland.	None.	21	"	"	Liquor Dealer.	Business.	"	1
413	46	"	U. S	Collegiate.	14	Fermented.	12 qts. champagne.	Lawyer.	Hereditary.	Fermented.	
414	47	Male.	U. S.	Com'n School.	27	Distilled.	1¼ pints.	Merchant.	Association.	Distilled.	1
415	54	"	"	Collegiate.	21	Both.	Largely.	None.	"	Both.	
416	58	"	"	Com'n School.	23	"	"	Merchant.	Loss of wife.	"	
417	45	"	"	None.	22	"	1 pint whiskey.	Laborer.	Association	"	
418	30	"	"	Com'n School.	21	Distilled.	1 pint.	Painter.	Hereditary.	Distilled.	
419	25	"	"	"	18	"	20 drinks.	Clerk.	Association.	"	
420	41	"	"	"	35	"	1½ pints.	Bookkeeper.	Domestic trouble.	"	
421	60	"	"	"	50	"	Largely.	Carpenter.	Association.	"	
422	22	"	"		20	Fermented.	3 quarts.	Farmer.	"	Fermented.	
423	29	"	"		22	Distilled.	1 pint.	Plasterer.	"	Distilled.	
424	40	"	"	"	20	"	1 quart.	"	"	"	1
425	34	"	"	"	23	Both.	Largely.	Insurance Ag't.	Business.	"	1

TABLE II—Continued.

Showing kind and quantity of drinks used by each of 500 Inebriates, treated in the KINGS COUNTY INEBRIATES' HOME at FORT HAMILTON, N. Y.

No.	Age.	Sex.	Place of Birth.	Education.	Age at which Inebriate began using intoxicants.	Kind of drink (fermented or distilled) used.	Maximum daily quantity that patient consumed of such drink.	Occupation when using intoxicants was commenced.	Mental or physical cause to which excesses are ascribed.	Kind of beverages to which necessity of med'l treatm'nt is attributed.	Number of attacks of delirium.
426	46	Male,	U. S.	Com'n School.	21	Distilled.	1 quart.	Clerk.	Association.	Distilled.	1
427	40	"	"	"	20	"	"	Engineer.	"	"	
428	45	"	England.	"	35	"	¾ pint.	Painter.	"	"	
429	61	"	Scotland.	"	30	"	1 quart.	Laborer.	"	"	1
430	38	"	U. S.	Collegiate.	27	"	½ pint to 3 quarts.	Clerk.	"	"	
431	42	"	"	Com'n School.	32	"	Largely.	Painter.	Hereditary.	"	
432	28	"	"	"	18	"	1 pint.	"	... .	"	
433	35	"	Ireland.	Collegiate.	27	"	1½ quarts.	Journalist.	Association.	"	
434	44	"	U. S.	Com'n School.	37	"	Period'l debauches.	Shoemaker.	"	"	8
435	35	"	"	"	25		2 to 20 drinks.	Clerk.	"	"	
436	59	"	Ireland.	None.	35	"	1 quart.	Teamster.	"	"	1
437	45	"	U. S.	Com'n School.	40	"	1½ pints.	Furrier.	"	"	
438	32	"	"	"	6	"	2½ quarts.	Stock raiser.	Hereditary.		
439	31	"	"	"	..	"	1½ pints.	Clerk.	Association.	"	
440	59	"	Ireland.	None.	50	"	"	Laborer.	"	"	
441	29	"	U. S.	Com'n School.	19	"	Largely.	Brakeman.	"	"	
442	38	Female.	Ireland.	"	34	"	1 pint.	Servant.	"	"	
443	39	"	"	"	28	"	1½ pint.	Gaiter Fitter.	Injury.	"	
444	29	Male.	U. S.	"	21	"	"	House Framer.	Association.	"	1
445	31	"	"	"	27	"	Largely.	Clerk.	"	"	
446	26	"	Scotland.	Collegiate.	23	"	1 pint.	"	Business trouble.	"	
447	38	Female.	Canada.	Com'n School.	34	"	½ pint.	None.	Hereditary.		
448	20	Male.	U. S.	"	19	"	1 quart.	"	Family trouble.	"	
449	43	"	"	Collegiate.	28	"	1½ quart.	Clerk.	Association.	"	
450	47	"	Ireland.	Com'n School.	42	"	......	Upholsterer.	"	"	
451	34	"	U. S.	"	28	"	1 pint.	Clerk.	Family trouble.	"	
452	61	"	"	Collegiate.	20	"	1½ pint.	Merchant.	Association.	"	
453	45	"	Porto Rico.	"	15	"	15 drinks,	None.	"	"	
454	41	"	U. S.	Com'n School.	24		......	Clerk.	"	"	2
455	58	"	"	"	28	"	Largely.	Compositor.	"	"	
456	25	"	"	"	23	Both.	½ pt. dist., also beer.	Canvasser.	"	Both.	
457	40	"	"	"	33	"	Largely.	Clerk.	Family trouble.	"	
458	30	"	"	"	15	Distilled.	1½ quart.	Farmer.	Association.	Distilled.	
459	51	"	"	"	41	"	Largely.	Carpenter.	Family trouble.	"	
460	26	"	Ireland.	"	18	Both.	"	Bartender.	Business.	"	

461	36	Male.	U. S.	Com'n School.	24	Distilled.	1¼ pint.	Waiter.	Association.	Distilled.	1
462	35	"	"	Collegiate.	20	"	Largely.	Contractor.	Nervous temp't.	"	
463	50	"	"	Com'n School.	20	"	1 quart.	Hatter.	Association.	"	
464	31	"	"	"	23	"	1 pint.	Salesman.	"	"	
465	35	"	"	"	13	"	Largely.	Hatter.	"	"	
466	54	"	"	"	49	"	1 pint.	Moulder.	Sickness.	"	
467	38	"	"	"	35	"	"	Harness Maker.	Family trouble.	"	
468	27	"	"	"	17	"	10 drinks.	Tinsmith.	Association.	"	
469	52	"	"	Collegiate.	27	Both.	Copiously.	Civil Engineer.	Grief.	Both.	
470	27	"	"	Com'n School.	16	Distilled.	Largely.	Hardware.	Business trouble.	Distilled.	
471	35	"	"	"	20	"	1 quart.	Hatter.	Association.	"	1
472	23	"	"	"	12	"	Largely.	None.	Hereditary.	"	
473	29	"	"	"	24	"	"	Engineer.	Association.	"	
474	29	"	"	"	28	"	1¼ pints.	None.	"	"	
475	27	"	"	"	21		1 pint.	Teamster.	"	"	
476	32	"	England.	Collegiate.	22	"	"	Grocer.	"	"	1
477	30	"	Scotland.	Com'n School.	26	"	Largely.	Agent.	"	"	
478	41	"	U. S.	"	38	"	1 pint.	Clerk.	"	"	
479	40	"	"	"	20	"	"	"	"	"	
480	22	"	Canada.	Collegiate.	20	"	1 quart.	"	"	"	
481	31	"	U. S.	Com'n School.	17	"	1¼ pints.	Druggist.	"	"	
482	55	"	"	Collegiate.	35	"	1 quart.	Broker.	"	"	3
483	41	"	England.	Com'n School.	31	"	1 pint.	Merchant.	"	"	
484	26	"	U. S.	"	23	"	Copiously at times.	Conductor.	"	"	
485	36	"	England.	"	24	"	10 drinks.	Adv'tising agt.	"	"	
486	21	"	U. S.	"	16	Fermented.	3 to 4 quarts beer.	None.	"	Fermented.	
487	26	"	"	"	24	Distilled.	Largely.	Clerk.	"	Distilled.	
488	25	"	"	"	20	"	1 qt. 5 consec. wks.	"	Grief	"	
489	24	"	"	"	22	"	10 drinks.	Laborer.	Association.	"	
490	30	"	"	Collegiate.	20	"	Largely.	Salesman.	Family trouble.	"	1
491	66	"	Germany.	Com'n School.	26	"	"	Shoemaker.	Association.	"	
492	47	"	U. S.	"	25	"	1 pint.	Cooper.	"	"	
493	38	Female.	"	"	22	"	"	None.	Hereditary.	"	2
494	43	Male.	"	Collegiate.	16	"	25 to 30 drinks.	Manufacturer.	Association.	"	1
495	52	"	"	Com'n School.	16	"	Largely.	Clerk.	"	"	
496	36	"	"	Collegiate.	26	"	1¼ pints.	Physician.	"	"	1
497	19	"	"	Com'n School.	18	"	....	Compositor.	"	"	
498	45	"	"	"	26	"	1 quart.	Salesman.	"	"	
499	44	"	"	"	34	Both.	Largely.	"	Family trouble.	"	
500	33	"	"	"	21	Distilled.	1 quart.	Merchant.	Hereditary.	"	

TABLE III.

Being a summary of Table II. in point of nativity, and kind of drinks used.

NATIVITY.									NUMBER OF INEBRIATES ADDICTED TO			
United States.	England.	Ireland.	Scotland.	Canada.	Germany.	Cuba.	Porto Rico.	TOTAL.	Distilled Liquors.	Distilled and Fermented Liquors.	Fermented Beverages.	TOTAL.
338	27	92	10	17	13	2	1	500	441	35	24	500

TABLE IV.

Showing age, sex, nativity, &c., of the twenty-four habitual consumers of fermented drinks, enumerated in Table II.

No.	Age.	Sex.	Birthplace.	Education.	Occupation.	Kind of Fermented Drinks.	Quantity Consumed Daily.	Complicating Disease.	Family History.	Number of attacks of delirium tremens.
1	48	Female.	Ireland.	None.	Domestic.	Ale and Lager.	2 quarts.	Concussion.	…………	
2	32	Male.	"	Collegiate.	Editor.	Beer.	1 quart.	…………	Grandfather ineb.	
3	38	"	United States.	Com'n School.	Clerk.	"	"	Hernia.	Father inebriate.	
4	26	"	England.	"	Artist.	Ale and Lager.	1 quart to 2 galls.	…………	…………	
5	22	"	United States.	Collegiate.	Lawyer.	Beer.	15 glasses.	…………	…………	
6	48	Female.	Ireland.	None.	Housekeeper.	"	Largely.	…………	Father inebriate.	
7	36	Male.	"	Rudimentary.	Merchant.	Wine and Beer.	1 quart.	…………	"	
8	23	Female.	United States.	Com'n School.	None.	Beer.	2 quarts.	Leucorrhœa.	"	
9	30	"	Germany.	"	Bartender.	Beer and Wine.	3 quarts beer.	…………	…………	1
10	20	Male.	United States.	"	Clerk.	Beer.	4 quarts beer.	Concussion.	…………	
11	23	"	"	"	"	"	1 quart to 1 gallon.	…………	…………	
12	29	"	"	"	Carpenter.	"	"	Syphilis.	…………	
13	45	Female.	Ireland.	None.	Servant.	"	2 quarts.	Concussion.	…………	
14	32	"	"	"	Housewife.	"	"	Disease gen'l organ.	…………	
15	47	Male.	United States.	Collegiate.	Merchant.	"	1 gallon.	Concussion.	…………	1
16	40	Female.	"	None.	Servant.	"	3 quarts.	…………	…………	
17	59	Male.	"	Com'n School.	Manufacturer.	"	"	Syphilis.	Father inebriate.	
18	53	"	Germany.	"	Broker.	"	Largely.	…………	…………	
19	50	Female.	United States.	None.	Housewife	"	2 quarts.	…………	…………	
20	56	"	"	Com'n School.	"	"	"	…………	…………	
21	23	Male.	"	Collegiate.	Hatter.	"	"	…………	…………	
22	26	"	"	"	Lawyer.	Wine.	12 qts. champagne.	Syph. & Pneumonia	F'r & Grandf'r ineb.	
23	22	"	"	Com'n School.	Farmer.	Beer.	3 quarts.	Epilepsy & Syphilis	…………	
24	21	"	"	"	None.	"	3 to 4 quarts.	Syphilis.	Father inebriate.	

TABLE V.

Being a summary of Table IV. in point of nativity and sex, quantity of fermented drinks consumed, and attacks of delirium.

Nativity.					Sex.			Number of Inebriates who Daily Consumed:						Number of Beer Drinkers having had Delirium.		
United States.	England.	Ireland.	Germany.	Total.	Male.	Female.	Total.	1 quart.	2 quarts.	3 quarts.	4 quarts.	Over 4 quarts.	Total.	Male.	Female.	Total.
15	1	6	2	24	15	9	24	3	8	4	5	4	24	1	1	2

APPENDIX C.

TABLES VI, VII AND VIII.

CAUSES OF DEPENDENCE OF SIX HUNDRED AND SEVENTY-ONE PAUPERS.

TABLE VI.

Showing Age, Nativity, Education, Occupation and Cause of Dependence of 671 Paupers, who are now, or were formerly, inmates of the Kings County Poorhouse.

No.	Age.	Nativity.	Education.	Former Occupation.	Family History.	Cause of Dependence.
1	38	Ireland	Read & Write	Laborer	Self Supporting.	Broken Leg.
2	57	Germany	"	Cook	"	Rheumatism.
3	30	"	"	Laborer	"	Lunacy.
4	55	Ireland	"	Engineer	"	Sickness.
5	54	Scotland	"	None	"	Paralysis.
6	36	Ireland	None	Laborer	"	Left Foot Mashed.
7	39	Jersey	"	Pedler	"	Loss of Left Arm.
8	46	Ireland	Read only	Laborer	"	Rheumatism.
9	40	"	"	"	"	Chills and Fever.
10	13	Penn	"	None	"	Want of Work.
11	65	Ireland	"	Laborer	"	Felon on Hand.
12	63	"	None	"	"	Want of Work.
13	23	N. Y.	"	Moulder	"	Burnt Foot.
14	34	Ireland	Read only	Laborer	"	Broken Shoulder.
15	37	"	"	"	"	Broken Arm & Thigh
16	52	Germany	"	"	"	Chills and Fever.
17	46	Ireland	None	"	"	General Debility.
18	65	Germany	"	"	"	Want of Work.
19	75	Ireland	Read only	"	"	Hurt to Left Eye.
20	45	"	"	"	"	Cripple.
21	26	Scotland	"	Sailor	"	Fever and Ague.
22	48	Ireland	None	Laborer	"	Want of Work.
23	45	"	"	"	"	Fever and Ague.
24	33	"	Read only	"	"	Vagrancy.
25	35	N. Y.	"	"	"	Malarial Fever.
26	40	England	"	Butcher	"	Boils on Left Leg.
27	53	Ireland	Read & Write	"	"	Chills and Fever.
28	47	N. Y.	None	Peddler	"	Amputated Leg.
29	34	Ireland	Read & Write	Laborer	"	Ulcer on Leg.
30	43	"	"	"	"	Bad Leg.
31	68	Germany	"	Farmer	"	Rheumatism.
32	71	"	"	Painter	"	"
33	46	Finland	Read only	Sailor	"	" Chronic.
34	13	N. Y.	None	None	Par'ts own house.	Want of Work.
35	50	Ireland	Read & Write	Carpenter	Self Supporting	Paralysis.
36	56	"	"	"	"	Piles & Gen'l debility.
37	31	"	None	Laborer	"	Want of Work.
38	57	Germany	"	Cabinetm'r.	"	Diarrhœa.
39	38	"	"	Currier	"	Neuralgia.
40	62	"	"	Turner	"	Asthma.
41	73	"	"	Tailor	"	Ruptured.
42	54	N. Y.	"	Car cond'r.	"	Syphilis.
43	76	Germany	"	Watch mkr.	"	Sickness.
44	20	N. Y.	"	Laborer	"	Want of Work.
45	38	Ireland	"	"	"	Chills and Fever.
46	42	"	Read & Write	"	"	Inflammatory Rhe'm.
47	35	Germany	"	Baker	"	Kidney Disease.
48	50	Ireland	"	Laborer	"	Dislocated Shoulder.
49	52	"	None	"	"	Injury to Knee.
50	65	Germany	Read & Write	"	"	Sickness.
51	53	"	"	Painter	"	Rheumatism.
52	21	N. Y.	"	Waiter	"	Sore Feet.
53	60	Delaware	None	Car driver	"	Want of Work.
54	58	Ireland	None	Laborer	"	Want of Work.
55	20	N. Y.	Read & Write	"	"	" "
56	71	"	"	"	"	Rheumatism.
57	41	England	"	"	"	Want of Work.
58	64	N. Y.	"	Painter	"	Ruptured.
59	51	Germany	"	Laborer	"	General Debility.
60	40	Ireland	Read only	"	"	Paralysis.
61	48	"	None	"	"	Sore Leg.
62	56	N. Y.	"	"	"	Vagrancy.
63	45	Ireland	Read only	"	"	Epileptic Fits.
64	40	"	Read & Write	Tailor	"	Sickness.
65	60	"	Read only	Laborer	"	Want of Work.
66	72	Germany	Read & Write	Painter	"	Paralysis.
67	45	Ireland	None	Laborer	"	Ulcers, stoppage of U.
68	38	"	Read & Write	"	"	Want of Work.
69	35	Germany	"	"	"	Ruptured.
70	64	Ireland	"	Engineer	"	Paralysis of Bladder.
71	67	Germany.	"	Laborer	"	Want of Work.
72	67	Ireland	Read only	"	"	Burnt Foot.
73	60	Germany	Read & Write	Gardener	"	Consumption.

TABLE VI—*Continued.*

Showing Age, Nativity, Education, Occupation and Cause of Dependence of 671 Paupers, who are now, or were formerly, inmates of the Kings County Poorhouse.

No.	Age.	Nativity.	Education.	Former Occupation.	Family History.	Cause of Dependence.
74	41	Germany	Read only	Laborer	Self Supporting	Want of Work.
75	14	N. Y.	"	None	"	"
76	36	Ireland	Read & Write	Laborer	"	"
77	40	"	Read only	"	"	Rheumatism.
78	43	"	Read & Write	"	"	Want of Work.
79	71	"	"	Basket mk.	"	Defective Sight.
80	28	"	"	Laborer	"	Rheumatism.
81	54	Germany	"	Segars	"	Paralysis.
82	40	"	"	Shoemaker.	"	Internally Injured.
83	52	Ireland	"	Tinsmith	"	Want of Work.
84	56	Germany	"	Laborer	"	Disjointed Foot.
85	51	"	"	"	"	Want of Work.
86	57	Ireland	"	"	"	Asthma.
87	32	"	"	"	"	Want of Work.
88	40	"	"	"	"	Loss of Leg.
89	53	"	"	"	"	Bad Cold.
90	72	Germany	"	"	"	Rheumatism.
91	40	Ireland	"	"	"	Left Leg Injured.
92	69	"	None	"	"	Want of Work.
93	50	N. Y.	"	"	"	Frostbitten & Cough.
94	68	Ireland	"	"	"	Partial Loss of Sight.
95	40	"	Read only	"	"	Chills and Fever.
96	70	Germany	Read & Write	Gardener	"	Age.
97	43	W. Indies.	"	Civil Eng	"	Rheumatism.
98	85	Ireland	None	Laborer	"	Age.
99	39	N. Y.	Read & Write	Car driver	"	Sore Leg.
100	23	"	"	Peddler	"	Rheumatism
101	70	Ireland	Read only	Carpet wvr.	"	Want of Work.
102	70	"	Read & Write	None	"	"
103	63	"	"	Blacksm'h	"	Bad Cold.
104	67	England	"	Bricklayer	"	Want of Work.
105	62	Germany	"	Farmer	"	Sickness.
106	76	Ireland	Read only	Laborer	"	Bad Health.
107	73	N. Y.	"	Carpet wvr.	"	Want of Work.
108	46	Ireland	Read & Write	Laborer	"	Disease of Kidneys.
109	63	Germany	"	Cook	"	"
110	47	Germany	Read & Write	Laborer	"	Lunacy.
111	37	Ireland	"	Peddler	"	Loss of Leg.
112	48	"	"	Machinist	"	Leg & Arm disabled.
113	34	Germany	"	Blacksmith	"	Sore Foot.
114	44	N. Y.	None	Peddler	"	Asthma, nearly Blind.
115	54	Germany	Read & Write	Gardener	"	Want of Work
116	55	Ireland	"	Cook	"	Hurt to Back & Head.
117	31	"	"	Laborer.	"	Loss of Leg.
118	30	"	None	"	"	Sore Leg.
119	47	N. Y.	Read & Write	Shoemaker.	"	Paralysis.
120	75	Ireland	"	Gardener	"	Rheumatism.
121	41	England	"	Sailor	"	Vagrancy.
122	65	Germany	"	Shoemaker.	"	Rheumatism.
123	55	"	"	"	"	"
124	66	Penna	"	Penmaker	"	Ruptured.
125	55	Germany	None	Laborer	"	Broken Leg.
126	33	N. Y.	Read & Write	Cook	"	Want of Work.
127	52	Ireland	None	Laborer	"	Loin of Back Broken.
128	38	Germany	Read & Write	"	"	Fever and Ague.
129	42	Ireland	None	"	"	Bad Cold.
130	40	"	Read & Write	"	"	Ulcers on both Legs.
131	59	Scotland	"	Bookk'per	"	Rheumatism, Rupt'd.
132	57	Germany	"	Chair cani'g	"	"
133	60	Ireland	"	Laborer	"	Bad Cold.
134	38	"	Read only	"	"	Rheumatism.
135	38	"	Read & Write	"	"	Chills and Fever.
136	75	England	"	"	"	Vagrancy.
137	58	Germany	Read only	"	"	"
138	87	N. Jersey.	None	None	"	Age.
139	54	Germany	Read & Write	Laborer	"	Rheumatism.
140	47	Ireland	"	None.	"	Broken Leg.
141	55	"	None	Laborer	"	Rheumatism.
142	50	"	"	"	"	"
143	35	Germany	Read & Write	Puddler.	"	Chills and Fever.
144	33	Ireland	None.	Laborer	"	Want of Work.
145	43	"	Read & Write	Porter	"	Rheumatism.
146	47	"	None	Laborer	"	Sore Foot.

TABLE VI—*Continued.*

Showing Age, Nativity, Education, Occupation and Cause of Dependence of 671 Paupers, who are now, or were formerly, inmates of the Kings County Poorhouse.

No.	Age.	Nativity.	Education.	Former Occupation.	Family History.	Cause of Dependence.
147	40	Ireland	Read only	Fireman	Self Supporting	Chills and Fever.
148	67	"	None	Laborer	"	Rheumatism.
149	64	"	Read & Write	"	"	Dropsy.
150	34	"	None	Shoemaker.	"	Want of Work.
151	45	"	Read & Write	Clockmkr	"	Consumption.
152	50	"	Read only	Laborer	"	Want of Work.
153	39	"	Read & Write	"	"	Broken Arm & Deaf.
154	65	Maryland	None	"	"	General Debility.
155	32	"	Read & Write	Waiter	"	Skin Disease.
156	35	Ireland	None	Teamster	"	Broken Kneecap.
157	68	Germany	Read & Write	Tailor	"	Rheumatism.
158	40	Ireland	None	Laborer	"	Want of Work.
159	56	"	Read & Write	Stonecut'r	"	Wounds in War.
160	32	Mass.	"	Painter	"	Heart Disease.
161	25	Germany	"	Plasterer	"	Sickness.
162	63	Ireland	"	Apple sta'd.	"	Want of Work.
163	28	"	"	Laborer	"	" "
164	43	"	"	"	"	Fractured Ribs.
165	35	"	None	Laborer	"	Heart Disease.
166	63	"	Read & Write	"	"	Want of Work.
167	39	"	"	None	"	Disability.
168	34	Germany	"	Farmer	"	Rheumatism.
169	63	"	"	"	"	"
170	23	N. Y.	"	Driver	"	Broken Collar Bone.
171	42	Ireland	"	Blacksmith	"	Disability.
172	52	"	None	Peddler	"	Blind.
173	43	"	"	"	"	Disability.
174	34	Scotland	Read & Write	Iron Workr	"	Rheumatism.
175	55	Ireland	"	Laborer	"	General Debility.
176	35	N. Y.	"	Peddler	"	Syphilis.
177	45	Germany	"	Ropemaker	"	Rheumatism.
178	46	Ireland	None	Laborer	"	Dyspepsia.
179	50	France	Read & Write	Painter	"	Partial Loss of Sight.
180	30	N. Y.	"	Fireman	"	Vagrancy.
181	67	Germany	"	Basketmkr.	"	Rheumatism.
182	73	"	None.	Gardener	"	Age.
183	14	N. Y.	"	None	"	Fits.
184	37	Ireland	"	Laborer	"	Bronchitis.
185	34	Germany	Read & Write	Wheelright	"	Disability.
186	45	Ireland	None	Laborer	"	Frostbitten.
187	65	"	"	"	"	Rheumatism.
188	33	"	"	"	"	Sickness.
189	36	"	Read & Write	"	"	Rheumatism.
190	30	Germany	"	File cutter.	"	Paralysis.
191	58	"	"	Shoemaker.	"	Sickness.
192	66	Ireland	None	Laborer	"	Partial Blindness.
193	26	"	Read & Write	"	"	Vagrancy.
194	38	N. Y.	"	"	"	Disability.
195	59	Italy	None	Ship carp'r.	"	Broken Leg.
196	36	Ireland	Read & Write	Plumber	"	Want of Work.
197	48	Scotland	"	Driver	"	" "
198	58	Ireland	"	Painter	"	Rheumatism.
199	75	Germany	"	Peddler	"	Vagrancy.
200	54	"	"	Blacksm'h.	"	"
201	65	"	"	Soapmaker.	"	"
202	50	Ireland	None	Laborer	"	Want of Work.
203	69	"	Read & Write	Carpenter	"	Disability.
204	51		"	Tailor	"	General Debility.
205	30	Virginia	"	Laborer	"	Blind.
206	59	Germany	"	Varnisher	"	Vagrancy.
207	56	Ireland	"	Tailor	"	Want of Work.
208	28	"	"	Laborer	"	Disability.
209	39	"	"	Tanner	"	Want of Work.
210	80	"	"	Peddler	"	Rheumatism.
211	31	"	"	Longsho'n.	"	Broken Leg.
212	62	Germany	"	Locksmith.	"	Broken Shoulder.
213	65	Ireland	None	Laborer	"	Want of Work.
214	18	N. Y.	Read & Write	Waiter	"	Fits.
215	17	Scotland	"	None	"	Inflam. of Lungs.
216	62	Ireland	"	Laborer	"	Want of Work.
217	64	Germany	"	Cutler	"	Disability.
218	22	N. Y	Read only	Cooper	"	Vagrancy.
219	50	Ireland	"	Laborer	"	Cripple.

TABLE VI—*Continued.*

Showing Age, Nativity, Education, Occupation and Cause of Dependence of 671 Paupers, who are now, or were formerly, inmates of the Kings County Poorhouse.

No.	Age.	Nativity.	Education.	Former Occupation.	Family History.	Cause of Dependence.
220	61	N. Y.	Read & Write	Soap maker	Self Supporting	Broken Arm.
221	67	Ireland	None	Laborer	"	General Debility.
222	55	Germany	Read & Write	Tailor	"	" "
223	46	N. Y.	"	Laborer	"	Chills and Fever.
224	63	Germany	"	Shoemaker.	"	Bronchitis.
225	68	"	"	Carpenter	"	Broken Leg & Rh'm.
226	80	"	"	Rope maker	"	Hemorrhage of Lungs
227	42	Ireland	"	Cartman	"	Fractured Shoulder.
228	61	"	"	Laborer	"	General Debility.
229	53	"	None	"	"	Chills and Fever.
230	48	N. Y	"	Blacksmith	"	Disability, sore legs.
231	60	Ireland	"	Laborer	"	" sore leg.
232	65	"	"	"	"	General Debility.
233	70	"	"	"	"	" "
234	63	England	Read & Write	Shoe mfr	"	Vagrancy.
235	20	N. Y	"	Printer	"	Deaf.
236	80	Ireland	None	Laborer	"	Age.
237	65	"	"	"	"	General Debility.
238	45	"	"	"	"	Disability, cut head.
239	15	N. Y.	"	None	"	Deaf, Dumb & Defor.
240	58	Ohio	Read only	Shoemaker.	"	Disability.
241	53	Ireland	"	Farmer	"	Rheumatism.
242	23	N. Y.	Read & Write	Printer	"	Consumption.
243	70	"	"	Plowmaker	"	Malaria.
244	13	Norway	Write only	None	"	Want of Work.
245	60	Ireland	None	Hod carrier	"	General Debility.
246	50	Germany	Read & Write	Tailor	"	Consumption.
247	78	"	None	Mason	"	Blind of left Eye.
248	56	N. Y.	"	Laborer	"	Fits.
249	46	Hungary	"	"	"	Disability.
250	60	Ireland	Read & Write	Tailor	"	General Debility.
251	70	Scotland	"	Blacksmith	"	Disability bad leg.
252	41	N. H	"	Laborer	"	Malaria.
253	67	Ireland	"	Gro. & Liq.	"	Rheumatism.
254	68	"	None	Gardener	"	General Debility.
255	76	Germany	Read & Write	None	"	Billious Fever.
256	55	N. Y.	"	Mariner	"	Sunstroke.
257	60	Ireland	"	Mason	"	Loss of right Leg.
258	37	Sweden	"	Laborer	"	Lung Disease.
259	59	Germany	"	Pipe maker	"	Rheumatism.
260	50	N. Y	"	Laborer	"	General Debility.
261	20	"	Read only	None	"	Paralysis.
262	45	Austria	"	Fireman	"	Rheumatism.
263	62	Ireland	"	Laborer	"	Disability.
264	16	N. Y	None	None	"	Paralysis.
265	76	Ireland	"	Laborer	"	Want of Work.
266	60	N. J.	Read & Write	Agent	"	Bronchitis.
267	70	England	"	Shoemaker.	"	Impaired Mind.
268	51	N. Y.	Read only	Laborer	"	Syphilis.
269	66	Germany	Read & Write	None	"	Partial Blindness.
270	65	Ireland	None	Laborer	"	Lunacy.
271	58	Germany	Read & Write	Blacksmith	"	Broken Hip.
272	70	"	"	Farmer	"	Ruptured.
273	75	Ireland	None	Laborer	"	General Debility.
274	26	Georgia	Read & Write	Druggist	"	" "
275	39	Ireland	"	Gardener	"	Disability.
276	78	N. Y.	None	Longshor'n	"	Rheumatism.
277	48	Ireland	"	Horseshoer.	"	Fractured Hip.
278	49	"	Read & Write	Bookkeep'r	"	Disability.
279	60	"	"	Laborer	"	Partial Blindness.
280	72	"	None	Junkman	"	Broken Ribs.
281	32	"	Read only	None	"	Epileptic Fits.
282	53	Mass.	Read & Write	Laborer	"	Disability
283	35	England	"	Machinist	"	Sickness.
284	58	Germany	"	Tailor	"	Rheumatism, Chronic
285	18	N. Y.	"	None	"	Want of Work.
286	65	Ireland	Read only	Laborer	"	Disjointed Hip.
287	68	Germany	Read & Write	Builder	"	Rheumatism, Chronic
288	25	N. Y.	None	None	"	Paralysis.
289	71	Germany	Read & Write	Laborer	"	Rupture and Debility.
290	38	Ireland	None	"	"	Disability.
291	62	N. Y	"	Tailor	"	Vagrancy.
292	63	Ireland	"	Baker	"	Paralysis.

TABLE VI—*Continued.*

Showing Age, Nativity, Education, Occupation and Cause of Dependence of 671 Paupers, who are now, or were formerly, inmates of the Kings County Poorhouse.

No.	Age.	Nativity.	Education.	Former Occupation.	Family History.	Cause of Dependence.
293	37	England	None	Laborer	Self Supporting	Blindness & Rupture.
294	43	Ireland	"	Soldier	"	Disability, Sore Foot.
295	80	Germany	"	Painter	"	Want of Work.
296	48	Ireland	"	Gardner	"	Asthma.
297	41	"	"	Carpenter	"	Want of Work.
298	45	"	"	Laborer	"	Blind and Crippled.
299	52	N. Y.	"	H'se Paint'r	"	Heart Dis. & Rheum.
300	57	Germany	"	Laborer	"	Ruptured.
301	71	N. Y.	"	Ship Carp'r	"	Sickness.
302	67	England	"	Hatter	"	Want of Work.
303	54	Ireland	"	Laborer	"	Syphilis and Rheum.
304	70	England	Read & Write.	"	"	Loss of both feet.
305	48	N. Y.	"	Carpenter	"	Lunacy.
306	67	France	"	Tailor	"	Disability.
307	65	Germany	"	Cabinet M'r	"	Sickness.
308	40	Ireland	None	Laborer	"	Debility.
309	54	"	"	Junkman	"	General Debility.
310	38	"	"	None	"	Partial Blindness.
311	42	"	"	Laborer	"	Consumption.
312	23	"	"	Barber	"	Sickness.
313	50	Germany	"	Carpenter	"	Want of Work.
314	35	Ireland	"	Laborer	"	Vagrancy.
315	40	"	"	"	"	Pneumonia.
316	42	"	Read only	"	"	Bad Cold.
317	59	N. Y.	"	Tin Smith	"	Loss of Foot.
318	74	West Va.	"	"	"	Vagrancy.
319	77	Ireland	None.	Laborer	"	Want of Work.
320	80	"	Read only	"	"	Heart and Lung Dis.
321	18	N. Y.	"	Farmer	"	Want of Work.
322	66	Ireland	"	Cooper	"	Disability.
323	50	"	"	Laborer	"	Broken Foot.
324	60	"	None	Blacksmith	"	Disability.
325	59	"	"	Stone Cut'r.	"	"
326	36	Germany	"	Laborer	"	Chills and Fever.
327	58	Ireland	"	Mat Maker.	"	Hip Disease.
328	41	"	"	Shoemaker	"	Nervous Debility.
329	60	"	"	Laborer	"	Lameness.
330	50	Germany	"	"	"	Chills and Fever.
331	67	N. Y.	"	Carpenter.	"	Broken Foot.
332	60	Ireland	None	Laborer	"	General Debility.
333	60	England	Read & Write	Carpenter	"	Disability.
334	60	Ireland	None	Laborer	"	Rheumatism.
335	44	Germany	"	"	"	Broken Leg.
336	39	"	"	Carpenter	"	Disability.
337	67	England	"	Clo'ng Cut'r	"	Ruptured.
338	64	Ireland	"	Painter.	"	Disability.
339	67	N. B'wick.	"	Boatman	"	Chills & Rheumatism.
340	27	England	"	Porter	"	Broken Leg.
341	50	N. Y	"	Laborer	"	Ruptured.
342	48	Penn.	"	None	"	Complicated Diseases
343	64	Ireland	"	Shoemaker.	"	Want of Work.
344	65	"	"	Laborer	"	" "
345	33	Sweden	"	"	"	Lung Disease.
346	46	Ireland	Read only	"	"	Lung & Heart Disease
347	79	Germany	"	"	"	General Debility.
348	40	Ireland	"	Waiter	"	Broken Ribs.
349	50	"	"	Laborer	"	Lung & Kidney Dis.
350	75	Germany	"	Tailor	"	Vagrancy.
351	53	Ireland	"	Farmer	"	"
352	64	Scotland	"	Tailor	"	Disability.
353	47	Germany	"	Laborer	"	"
354	34	N. Y.	"	"	"	Lameness.
355	52	"	"	"	"	Deaf and Ruptured.
356	48	Germany	None	Farm Hand	"	Lameness.
357	63	S. Carolina	"	Hostler	"	Asthma.
358	63	Ireland	"	Laborer	"	Disability.
359	59	Prussia	Read & Write	Laborer	"	Disability.
360	58	N. Y.	"	Tinsmith	"	Vagrancy.
361	64	Ireland	None	None	"	Paralysis.
362	37	Germany	Read & Write	Farm Hand	"	Vagrancy.
363	64	N. Y	Read only	Mason	"	Lameness.
364	58	England	Read & Write	Blacksm'h	"	Malaria.
365	53	Ireland	"	Laborer	"	Want of Work.

TABLE VI—*Continued.*

Showing Age, Nativity, Education, Occupation and Cause of Dependence of 671 Paupers, who are now, or were formerly, inmates of the Kings County Poorhouse.

No.	Age.	Nativity.	Education.	Former Occupation.	Family History.	Cause of Dependence.
366	65	"	None	Laborer	Self Supporting	" "
367	40	N. Y.	"	"	"	Sickness.
368	47	Ireland	"	"	"	Vagrancy.
369	61	Germany	Read & Write	Shoemaker.	"	Lameness.
370	42	Ireland	"	Carpenter	"	Spinal Disease.
371	43	N. Y.	"	Iron Railer.	"	Lunacy.
372	35	Germany	None	Farm Hand	"	Want of Work.
373	55	Ireland	"	Laborer	"	Partial Blindness.
374	63	"	"	"	"	Want of Work.
375	58	"	Read & Write	Canvasser	"	Dest. and Homeless.
376	60	Germany	"	Tailor	"	Want of Work.
377	36	Ireland	"	Cook	"	Heart Disease.
378	19	N. Y.	"	Stonecutter	"	Want of Work.
379	27	Ireland	"	Laborer	"	" "
380	68	N. Y.	"	Shoemaker	"	Disability.
381	47	Main	"	Salesman	"	Want of Work.
382	40	Sweden	"	Iron F'ndry	"	" "
383	38	Ireland	"	Hostler	"	Disability.
384	42	"	"	Laborer	"	"
385	53	Canada, W	"	Printer	"	"
386	61	Ireland	"	Laborer	"	Lameness.
387	66	N. Jersey	None.	Unknown	"	Disability.
388	55	Ireland	Read & Write	Laborer	"	Lameness.
389	47	"	"	"	"	Vagrancy.
390	38	Germany	"	Tailor	"	Epileptic Fits.
391	39	N. Y.	"	Painter	"	Rheumatism.
392	25	N. Y.	None	Waiter.	"	Disability.
393	66	Germany	"	Butcher	"	"
394	65	Ireland	"	Laborer	"	Want of Work.
395	45	"	"	"	"	Disability.
396	38	"	"	Clerk	"	"
397	30	Sweden	"	Cigarmaker	"	Lunacy.
398	54	Ireland	Read only	Laborer	"	Disability.
399	31	"	Read & Write	Laborer	"	"
400	69	England	"	Engineer	"	Vagrancy.
401	64	"	"	Blacksm'h	"	Disability.
402	67	"	"	Iron M'lder	"	Partial Blindness.
403	70	"	"	Bookbinder	"	Cripple.
404	55	Germany	"	Tool Maker	"	Disability.
405	42	"	"	Peddler	"	Vagrancy.
406	22	N. Y.	"	Laborer	"	Loss of Leg.
407	50	Germany	"	Farm Hand	"	Want of Work.
408	54	"	"	Gardener	"	" "
409	60	Ireland	"	Laborer	"	Consumption.
410	40	Germany	"	"	"	Disability.
411	29	Ireland	"	"	"	Dislocated Arm.
412	81	Germany	None	Peddler	"	Vagrancy.
413	42	"	"	Farm Hand	"	Disability.
414	62	Ireland	"	Laborer	"	Want of Work.
415	56	Ireland	Read & Write	Nail Maker.	"	Disability.
416	35	"	"	Cofe Roas'r	"	"
417	23	"	"	Laborer	"	Sypbilis.
418	43	England	"	Painter	"	Colic.
419	32	Ireland	"	Laborer	"	Sickness.
420	47	"	None	Pedler	"	Loss of Leg.
421	26	"	"	Laborer	"	Want of Work.
422	44	N. Y.	"	"	"	Disability.
423	40	Ireland	"	Gardener	"	Want of Work.
424	68	Maine	Read only	Tailor	"	Disability.
425	35	Ireland	Read & Write	Painter.	"	General Debility.
426	30	N. Y.	"	Laborer	"	Loss of Left Hand.
427	45	"	Read only	"	"	Want of Work.
428	30	Ireland	None	Farm Hand	"	Vagrancy.
429	54	"	"	Laborer	"	Disability.
430	50	England	Read & Write	Painter	"	Want of Work.
431	30	Ireland	"	Box Maker.	"	Lung Disease.
432	48	"	None	Laborer	"	Disability.
433	57	Germany.	Read & Write	Basket Mkr.	"	Want of Work.
434	65	Ireland	"	Carpenter	"	Disability.
435	58	"	"	"	"	Want of Work.
436	30	"	Read only	Laborer	"	Sypbilis.
437	50	"	None.	"	"	Disability.
438	65	"	Read & Write	Shoemaker.	"	Sore Eyes.

TABLE VI—*Continued.*

Showing Age, Nativity, Education, Occupation and Cause of Dependence of 671 Paupers, who are now, or were formerly, inmates of the Kings County Poorhouse.

No.	Age.	Nativity.	Education.	Former Occupation.	Family History.	Cause of Dependence.
439	42	Ireland....	Read & Write..	Machinist..	Self Supporting..	Want of Work.
440	49	England...	"	Laborer....	"	" "
441	25	Germany..	None........	None.......	"	Idiocy.
442	60	Ireland....	Read & Write..	Laborer....	"	Disability.
443	70	N. Y....	"	"	"	General Debility.
444	45	Ireland....	Read & Write..	Laborer....	"	Internal Injury.
445	60	Scotland..	"	"	"	Want of Work.
446	35	N. J.......	None	None......	"	Sick.
447	64	Ireland....	"	Tin worker.	"	Scrofula.
448	56	"	"	None...	"	Vagrancy.
449	50	"	Read & Write..	Bricklayer.	"	Rheumatism.
450	60	"	Read only.....	Cobbler.....	"	General Debility.
451	58	Penn......	Read & Write..	Salesman..	"	Vagrancy.
452	52	Ireland....	None	Laborer....	"	Chills and Fever.
453	45	"	"	"	"	Want of Work.
454	75	"	Read & Write..	"	"	Rupt., Rheu., & Age.
455	42	"	"	"	"	Malaria and Rheu.
456	60	"	None	Cook.......	"	Sore Leg.
457	46	"	Read & Write..	Laborer....	"	Chills, FevMal&Rheu.
458	38	Germany..	"	Baker	"	Rheumatism.
459	70	Ireland....	"	Laborer....	"	General Debility.
460	77	"	None	"	"	Rheumatism and Age
461	47	"	"	"	"	Want of Work.
462	68	"	Read only.. ..	Harnessm'r	"	Age.
463	39	N. Y.. ...	"	Laborer	"	Lunacy.
464	44	Norway...	"	"	"	Paralysis.
465	50	Ireland....	"	"	"	Loss of Toe.
466	52	"	Read & Write..	Baker..	"	Sore Leg.
467	50	"	"	Laborer. ..	"	Arm & Legs Broken.
468	57	Germany..	"	"	"	Want of Work.
469	58	"	"	Shoemaker.	"	Kidney Disease.
470	17	Holland...	"	Farmer	"	Vagrancy.
471	18	N. Y......	"	Laborer ...	"	Want of Work.
472	62	N. Y	Read & Write..	Butcher ...	"	Want of Work.
473	69	Ireland....	"	Cook.......	"	Rheumatism & Cough
474	28	N. Y......	"	None	"	Paralysis.
475	46	"	"	Fireman....	"	Pan of Knee Broken.
476	44	Australia.	"	Book Ag'nt	"	Want of Work.
477	54	Ireland....	"	Laborer ...	"	Sore Leg.
478	70	"	"	None	"	General Debility.
479	28	N. Y.....	"	Laborer....	"	Swelled Foot.
480	56	Ireland....	None.........	Tailor	"	Paralysis.
481	53	"	Read & Write..	None	"	Sick.
482	75	England ..	Read only......	Cook.......	"	Swelled ft., age, des'n
483	44	Ireland....	None	Laborer....	"	Hurt to Right Knee.
484	76	N. Y......	"	None	"	Deaf.
485	60	England ..	Read & Write..	Engineer ..	"	Gunshot w'nd in Leg.
486	66	Ireland ..	None	Laborer....	"	Ruptured.
487	73	Scotland...	Read & Write..	Furrier ...	"	Want of Work.
488	55	Germany..	"	Blacksmith	"	Diarrhœa.
489	59	Ireland....	"	Laborer....	"	Sunstroke, kid. & lung
490	76	England ..	"	None	"	Paralysis.
491	35	Ireland....	"	Engineer ..	"	Contus'n on face & bk.
492	58	Germany..	"	Laborer ...	"	Bad Health.
493	31	Ireland....	"	Shoemaker.	"	Vagrancy.
494	62	Scotland..	"	Laborer....	"	Hurt to Head.
495	42	Germany..	"	Tailor. ...	"	Rheumatism.
496	50	"	"	Machinist .	"	Softening of Brain.
497	70	N. Jersey.	"	Flagger	"	Paralysis.
498	63	Ireland....	"	Laborer....	"	Want of Work.
499	60	"	"	"	"	General Debility.
500	67	Ireland....	"	Laborer....	"	Sore Leg.
501	54	Denmark	"	Shoemaker.	"	Kidney Dis. & Gravel.
502	63	Ireland....	None	Laborer....	"	Sunstroke.
503	26	England ..	Read & Write..	"	"	Want of Work.
504	73	Germany..	"	Gardener ..	"	Gen'l Debility & Age.
505	63	Ireland....	Read only......	Laborer....	"	Paralysis.
506	60	Germany..	None	"	"	Sore Feet.
507	70	"	"	"	"	Bad Leg.
508	65	"	Read & Write..	Fish Ped'ler	"	Rheumatism.
509	58	Ireland....	"	Laborer....	"	Inflammatory Rheu'm
510	57	Germany..	"	Barber.....	"	Running Sore in Leg.
511	60	Ireland....	"	Tailor......	"	Rheumatism.

TABLE VI—*Continued.*

Showing Age, Nativity, Education, Occupation and Cause of Dependence of 671 Paupers, who are now, or were formerly, inmates of the Kings County Poorhouse.

No.	Age.	Nativity.	Education.	Former Occupation.	Family History.	Cause of Dependence.
512	66	Ireland	Read & Write	"	Self Supporting	General Debility.
513	72	Norway	None	Watchman.	"	Want of Work.
514	74	Ireland	"	Laborer	"	Sore Leg.
515	71	"	Read & Write	"	"	Want of Work.
516	30	N. Y.	None	Carver	"	Deaf and Dumb.
517	52	Switzer'd.	Read & Write	Tailor	"	Rheumatism.
518	63	Ireland	"	Laborer	"	Neuralgia & Rheum'm
519	24	N. Y.	Read only	"	"	Imbecility of Mind.
520	66	Ireland	None	"	"	Blind.
521	72	"	"	Cartman	"	Want of Work.
522	60	"	Read & Write	Laborer	"	Frost Bitten.
523	64	"	"	"	"	Hurt to Body.
524	57	"	"	"	"	Consumption.
525	46	Germany	"	"	"	Deformed of Body.
526	60	Ireland	None	"	"	Leg and Arm Broken.
527	79	Germany	Read & Write	Farmer	"	Rheumatism, Old Age
528	70	Ireland	"	Carpet Wvr.	"	Bad Health.
529	80	"	None	Laborer	"	Age.
530	54	"	"	None	"	Paralysis.
531	55	"	Read & Write	Stone Mas'n	"	Skin Disease.
532	48	Germany	"	Blacksmith	"	Vagrancy.
533	20	Ireland	"	None	"	Bad Eyes.
534	69	"	None	Laborer	"	Sickness.
535	25	Italy	"	"	"	"
536	40	England	Read & Write	Coachman	"	Paralysis.
537	70	Ireland	"	Laborer	"	Rheumatism.
538	62	England	"	"	"	Ruptured.
539	38	Ireland	"	"	"	Pain in Left Side.
540	66	Germany	"	Cabinet m'r	"	Sore Foot and Cough.
541	56	Ireland	"	Shoemaker.	"	Partially Blind.
542	67	"	None	Laborer	"	Want of Work.
543	67	"	Read & Write	Shoemaker.	"	" "
544	61	Germany	"	"	"	Rheumatism.
545	35	N. Y.	"	Laborer	"	Vagrancy.
546	52	Virginia	None	"	"	Lunacy.
547	57	N. Y.	Read only	Gardener	"	"
548	45	Ireland	Read & Write	Laborer	"	Rheumatic Pains.
549	35	N. Y.	"	Lather	"	Lunacy.
550	74	Penn.	"	Oysterman.	"	Erysipelas.
551	22	"	None	Peddler	"	Paralysis.
552	76	Ireland	"	Laborer.	"	General Debility.
553	50	"	"	Servant	"	Want of Work.
554	50	"	Read & Write	Laborer	"	Erysipelas.
555	59	England	"	Organist	"	Diarrhœa.
556	55	Sweden	Read & Write	Laborer	"	Fever.
557	64	Ireland	"	"	"	Want of Work.
558	49	France	"	Glass Blow'r	"	Severe Cold.
559	57	Germany	"	Tailor	"	Severe Cold & Rheu'm
560	70	Ireland	"	Laborer	"	Left Hip Sprained.
561	28	Michigan	"	Cigar P'ker	"	Epileptic Fits.
562	65	Ireland	None	Laborer	"	Want of Work.
563	69	Germany	Read & Write	Seaman	"	" "
564	19	N. Y.	"	Carpenter	"	" "
565	54	Ireland	"	Laborer	"	Hurt to Breast & Rup.
566	65	England	"	Gold Beater	"	Rheum'sm & Sore Leg
567	71	Ireland	"	Shoemaker.	"	Want of Work.
568	33	"	"	Laborer	"	" "
569	37	Germany	"	"	"	" "
570	50	Norway	"	Tailor	"	Rheu'm and Debility.
571	63	Germany	"	Farmer	"	Impaired Mind.
572	50	Ireland	None	Laborer	"	Rheumatism.
573	24	N. Y.	Read only	Shoemaker.	"	Dropsy & Lung Dis'e.
574	80	Ireland	Read & Write	Laborer	"	Sore Leg.
575	64	Prussia	"	Dyer	"	Both Legs Broken.
576	59	Germany	"	Laborer	"	Sore Leg.
577	68	England	"	Tailor	"	Paralysis.
578	47	Ireland	Read only	Laborer	"	Sprained Arm.
579	45	N. Y.	Read & Write	Sailor	"	Rheumatism.
580	45	Ireland	"	Laborer	"	"
581	43	Canada	"	Porter	"	Want of Work.
582	37	N. Y.	"	Laborer	"	Loss of Right Leg.
583	40	Ireland	Read only	"	"	Right Leg Sprained.
584	14	N. Y.	None	None	Unknown	Vagrancy.

TABLE VI—*Continued.*

Showing Age, Nativity, Education, Occupation and Cause of Dependence of 671 Paupers, who are now, or were formerly, inmates of the Kings County Poorhouse.

No.	Age.	Nativity.	Education.	Former Occupation.	Family History.	Cause of Dependence.
585	30	N. Y.	Read & Write	Laborer	Self Supporting	Severe Cold on Lungs.
586	17	"	"	"	"	Want of Work.
587	36	Ireland	"	"	"	Sprained Hand & Foot
588	50	"	None	"	"	Lameness.
589	40	N. Y.	Read & Write	Tinsmith	"	Lunacy.
590	29	England	"	"	"	Epileptic Fits.
591	33	Ireland	"	Laborer	"	Rheum'sm & Asthma.
592	40	Holland	"	Clerk	"	Loss of Right Arm.
593	66	Germany	"	Engineer	"	Failing Sight.
594	38	Ireland	"	Laborer	"	Sprained Wrist.
595	34	"	None	"	"	Abscess in Neck.
596	47	England	Read & Write	"	"	Vagrancy.
597	35	N. Y.	None	"	"	Want of Work.
598	77	Ireland	"	"	"	Age.
599	35	Scotland	Read & Write	Moulder	"	Ulcer in Right Leg.
600	53	Ireland	Rudimentary	Lastmaker	"	Intemperance.
601	69	"	"	Laborer	"	"
602	31	"	"	"	"	"
603	44	"	"	None	"	"
604	27	Germany	"	Laborer	"	"
605	60	N. Y.	"	Penclcse mr	"	"
606	48	"	"	Cl'k & Bk'r	"	"
607	49	England	"	Farrier	"	"
608	50	"	"	Laborer	"	"
609	43	N. Y.	"	"	"	"
610	43	N. J.	"	Hatter	"	"
611	56	N. Y.	"	Laborer	"	"
612	34	"	None	"	"	"
613	62	"	Rudimentary	"	"	"
614	27	"	"	"	"	"
615	23	"	"	"	"	"
616	49	"	"	"	"	"
617	31	Ireland	Read only	Waiter	"	"
618	48	"	Rudimentary	Stonecutter	"	"
619	40	"	"	Laborer	"	"
620	71	N. Y.	"	Barber	"	"
621	40	"	"	Peddler	"	"
622	32	Ireland	"	Barber	"	"
623	55	"	None	Laborer	"	"
624	34	N. Y.	Rudimentary	Sailor	"	"
625	48	Ireland	"	Shoemaker	"	.
626	50	Ireland	"	Shoemaker	"	
627	65	Conn.	"	Ship Joiner	"	"
628	54	Ireland	"	Fireman	"	"
629	48	"	"	Laborer	"	"
630	31	"	None	"	"	"
631	27	N. Y.	"	Baker	"	"
632	45	"	Rudimentary	Car Cond'or	"	"
633	29	"	"	"	"	"
634	35	Ireland	"	Junkman	"	"
635	50	"	"	Bricklayer	"	"
636	30	"	Read only	Laborer	"	"
637	45	"	None	"	"	"
638	35	N. Y.	Rudimentary	Painter	"	"
639	55	Scotland	"	Tailor	"	"
640	69	Germany	"	Shoemaker	"	"
641	45	Ireland	"	Laborer	"	"
642	55	"	"	"	"	"
643	45	"	"	"	"	"
644	21	N. Y	"	"	"	"
645	39	Ireland	"	Boiler M'kr.	"	"
646	29	N. Y.		Laborer	"	"
647	32	"	None	"	"	"
648	43	Ireland	Rudimentary	"	"	"
649	69	"	"	Machinist	"	"
650	50	Penn	"	Tailor	"	"
651	70	Ireland	"	Sailmaker	"	"
652	43	N. Y.	"	Salesman	"	"
653	61	Ireland	"	Carpenter	"	"
654	38	Ireland	"	Engineer	"	"
655	55	Conn	"	Musician	"	"
656	32	Ireland	"	Tailor	"	"
657	66	"				

TABLE VI—*Continued.*

Showing Age, Nativity, Education, Occupation and Cause of Dependence of 671 Paupers, who are now, or were formerly, inmates of the Kings County Poorhouse.

No.	Age.	Nativity.	Education.	Former Occupation.	Family History.	Cause of Dependence.
658	38	Germany	Rudimentary	Tailor	Self Supporting	Intemperance.
659	42	Ireland	"	Laborer	"	"
660	49	"	"	Shoemaker.	"	"
661	59	"	"	Clerk.	"	"
662	55	"	"	Carpenter	"	"
663	60	"	"	Laborer	"	"
664	54	"	"	"	"	"
665	32	N Y	"	"	"	"
666	37	Ireland	Read only	"	"	"
667	41	N. Y	Rudimentary	Shipwright.	"	"
668	50	Ireland	"	Junkman	"	"
669	31	"	"	Housep'nt'r	"	"
670	21	N. Y	"	Painter	"	"
671	28	"	"	Locksmith.	"	"

TABLE VII.

Being Summary of Table VI. in point of Nativity, Age and cause of Indigence.

Number of Paupers.	Nativity.																		Causes of Dependence.					Age.							
	United States.	Ireland.	Germany.	England.	Scotland.	Sweden.	Norway.	France.	Italy.	Holland.	Canada.	Switzerland.	Denmark.	Austria.	Australia.	Hungary.	Finland.	West Indies.	Physical Disability.	Want of Work.	Vagrancy.	Age.	Intemperance.	From 10 to 20.	From 20 to 30.	From 30 to 40.	From 40 to 50.	From 50 to 60.	From 60 to 70.	From 70 to 80.	From 80 to 90.
671	142	332	124	35	13	5	4	3	2	2	2	1	1	1	1	1	1	1	457	99	33	10	72	17	47	112	134	143	146	62	10

TABLE VIII.

Showing nativity and age of the seventy-two paupers, whose dependence is attributable to intemperance.

Total Number of Paupers.	Nativity.					Age.					
	United States.	Ireland.	Germany.	England.	Scotland.	From 20 to 30.	From 30 to 40.	From 40 to 50.	From 50 to 60.	From 60 to 70.	From 70 to 80.
72	28	38	3	2	1	9	17	20	15	9	2

www.ingramcontent.com/pod-product-compliance
Lightning Source LLC
Chambersburg PA
CBHW020007030726
47500CB00002B/480

* 9 7 8 3 3 3 7 3 7 1 4 3 2 *